Balloons, Blériots and Barnstormers

200 YEARS OF FLYING FOR FUN

Balloons, Blériots and Barnstormers

200 YEARS OF FLYING FOR FUN

Alastair Goodrum

The History Press

First published 2009

The History Press
The Mill, Brimscombe Port
Stroud, Gloucestershire, GL5 2QG
www.thehistorypress.co.uk

British Library Cataloguing in Publication Data.
A catalogue record for this book is available from the British Library.

ISBN 978 0 7524 4516 8

Typesetting and origination by The History Press
Printed in Great Britain

Contents

Introduction

When the Marquis d'Arlandes and Pilatre de Rozier made the first manned free-balloon flight in 1783, the event was witnessed by thousands of people. Since time immemorial ordinary people have flocked to gaze upon daring feats – from the gladiators of ancient Rome to the intrepid aeronauts and aviators of modern times – in the hope of witnessing some death-defying or even death-inducing act in their very presence.

Aeronautics was both turned into a public spectacle and exploited financially from its very beginning and thus it has remained ever since. *Balloons, Blériots and Barnstormers* sets out to elaborate upon those concepts in the East Midlands and by doing so demonstrates that this aviation-rich region also encapsulates the story of how civil flying developed in the UK as an entertainment spectacle over two centuries.

The book achieves this bold claim successfully for three reasons:

Firstly, the sky above the East Midlands has never been quiet during the past 200 years, and life here has been involved with or touched by some form of aviation since the very birth of that branch of science. This eastern region as much as anywhere is a truly representative microcosm of the progress of general aviation in the whole of the United Kingdom.

Secondly, by focussing attention at a high level of detail and with such a large amount of human interest content, all of which occurs within a relatively small geographic area, the book enables the reader not only to grasp what is happening and why but also to feel the euphoria, frustrations and tensions of the airmen and airwomen involved, and experience the same feeling of excitement as the people who actually witnessed their efforts.

Thirdly, by taking the reader along a logical timeline of events, not only is a coherent reflection of the progress of civil general aviation created, but the reader can thoroughly 'enjoy the flight' too.

Ride the wind to dizzying heights with the most famous ballooning pioneers; jump for your life with naughty-nineties parachute girls; dice with death round pylons then dive for the winning line with daring air racers; climb out of the cockpit to wing walk with airmen with nerves of steel; swoop, dive, roll and loop the loop in formation with world famous record-breaking pilots. Yes, read on and become part of the show!

CHAPTER ONE

Up Where the Wind is Silent

James Sadler made his twenty-first ascent by balloon on 7 October 1811 and brought aeronautics to Lincolnshire in fine style. Riding on the back of a tempestuous wind, his journey became an epic of its time, being accomplished at a speed which remained unequalled until surpassed by an aeroplane more than 100 years later.

At 2 p.m. that day, in the Vauxhall district of Birmingham, the scene was set for this drama. Sadler's crimson and gold-striped balloon, 40ft tall and 36ft in diameter, was fully inflated with hydrogen gas – a time-consuming process which had begun the previous day.

The balloon, publicly displayed for the first time in Cambridge, presented a kaleidoscope of rich colours, its fabric, 3,632sq.ft of the finest silk, shimmering in the weak sunlight. Around its girth were three horizontal bands of different colours, one carrying the following inscription:

Celciss, Princeps Gulieimus Fredericus, Dux Gloucestrix, Acad. Cantab, Cancellarius Electus MDCCCXI.
(Roughly translated as, His Highness, Prince George Frederick, Duke of Gloucester, elected Chancellor of Cambridge University 1811)

Fully inflated, the balloon contained 20,500cu.ft of gas and the structure and contents weighed 1,286lb in total. To minimise the escape of gas through the fabric it was completely varnished on the interior side with a solution of Indian rubber. Suspended beneath the balloon, the car itself was a splendid sight to behold. It, too, was richly ornamented in a style that everyone agreed did Mr Sadler a great credit, having spared no expense to honour the illustrious prince whose name adorned the fabric. The car, 11ft long, 4ft wide and 3ft deep, was constructed of wickerwork covered with azure-coloured figured silk, studded with silver stars and made all the more striking by the addition of wreaths of laurel leaves arranged along the bottom. Six gilt ropes secured it below the balloon to a hoop adorned by a drapery of crimson satin, trimmed with golden acorns and bordered round the top with a wreath of oak leaves cut from gold foil. The drapery was secured to the hoop by loops of golden twist, and regency plumes were fixed to the seams of the car as two small silk flags fluttered in the breeze. It was a truly magnificent spectacle to behold.

Then, 'amidst an immense concourse of spectators', the veteran aeronaut and his friend John Burcham, from Swaffham in Norfolk, took their places in the open carriage beneath the 'aerostat' (balloon). At 2.20 p.m., final restraining ropes were cast off and the crimson

A VIEW OF THE BALLOON of Mr SADLERS ASCENDING
With him and Captain Paget of the Royal Navy from the Gardens of the Mermaid Tavern at Hackney on Monday Aug 12 1811.

James Sadler (left) and Captain Paget RN (right) taking off from the Mermaid Inn, Hackney, London on 12 August 1811, in the balloon in which Sadler and Burcham flew from Birmingham to Lincolnshire on 7 October 1811. (Spalding Gentlemen's Society)

and gold silk balloon rose majestically into the sky. Gaining height rapidly and heading east, after a few minutes it was enveloped by clouds and lost to view.

Born in Oxford, James Sadler lived from 1751 to 1828 and though he was a confectioner by trade – having inherited his father's business – he acquired a keen interest in science and was an ingenious inventor, a combination of abilities that eventually led him to the distinction of becoming the first English-born aeronaut. The term 'aeronaut' (navigator of the air) became the accepted term to describe someone who indulged in the new science of 'aerostation' (lighter-than-air flight) using an aerostat(ic) machine (balloon). He seems to have led an unremarkable early life prior to his first ascent, made in a hot-air balloon of the Montgolfier type in October 1784, but between that date and his last flight in 1816 he certainly made up for it.

While conducting his own early trial-and-error aeronautical experiments, first with heated air and then hydrogen, he became associated with Vincente Lunardi shortly after Lunardi made his historic first balloon ascent in England in 1784. During this period Sadler earned a reputation as an intrepid aerial adventurer that outstripped his contemporaries and in whose ballooning exploits no less a person than King George III, it was said, took a special interest. Having gained valuable experience with hydrogen gas while working with Lunardi, it was in May 1785 that Sadler became the first Englishman to use that lifting agent, which was originally named 'inflammable air', to fill a balloon.

James Sadler Esq.

First English Aeronaut.

Engraved for the Dublin Magazine.

James Sadler of Oxford
(1751–1828), the first English-
born aeronaut. (National
Library of Ireland)

This aptly named gas, discovered in 1766 by the English chemist Henry Cavendish, had that potentially lethal combination of both incendiary and lifting characteristics. Among its properties, Cavendish discovered, its specific gravity was at least seven to ten times lighter than atmospheric air. Hydrogen is the lightest of all the elements and 35cu.ft of the gas (1cu.metre) at 32°F (0°C) weighs just 0.242lb (0.11kg). On the downside, with the addition of as little as 6 per cent of ordinary air, hydrogen becomes a potentially explosive mixture. Cavendish, however, seems not to have pursued the former line of discovery along its logical path but there is anecdotal evidence that a contemporary of his, Dr Joseph Black, professor of chemistry at Glasgow University, first demonstrated the ability of hydrogen gas as a lifting agent. Henry Cavendish's work, though, provided the first catalyst leading to manned flight in lighter-than-air devices.

Black, too, seems to have ignored the further possibilities arising from his demonstration and it was not until 1782 in France that the Montgolfier brothers – who worked in their father's paper manufacture business – heard about, began to study and then applied theories expounded by Cavendish and another contemporary, Dr Priestley, about the specific gravities of different kinds of air. As a result, the second crucial event – and their fundamental contribution to manned flight – was their conclusion that by enclosing in a bag a quantity of gas that was lighter than its surrounding atmosphere, a weight might be lifted from the earth into the air. Although their initial experiments employed hydrogen

The eminent eighteenth-century scientist Dr Joseph Black was said to have first demonstrated the lifting capability of hydrogen gas. (Via Brian Cocks)

gas in paper and card containers, they were dissatisfied with the results mainly because the gas escaped from cracks and through the porous paper itself – not, in fact, because the gas itself was inappropriate. This disillusionment had the effect of diverting the brothers' attention from hydrogen. One of their experiments showed that ordinary air when heated to 180°F loses half its weight so they turned their attention to experiments with heated air with which, for all its inherent limitations, they managed to achieve the result they sought on 21 November 1783, and the rest, as they say, is history.

Inflammable gas became more universally referred to as hydrogen around 1790 and ousted hot air as the preferred lifting agent for aeronauts until the 1820s. Air, heated in the Montgolfier style, would lift about 15lb for every 1,000cu.ft of volume, while the same volume of hydrogen could lift about 70lb of payload. This meant that, although hydrogen was far more expensive, time consuming to produce and frequently of inconsistent quality, it was more efficient. Crucially, the lifting effect of hydrogen gas remained intact a great deal longer compared with the (sometimes alarming) rate at which lift was lost as heated air cooled, thus attaching to the former its own key contribution to the world of aeronautics – that of allowing longer, sustained flights to be made over greater distances.

It will be seen later that hydrogen, from 1821, would gradually give way to the use of coal gas as the preferred lifting agent for ballooning. Although still utilised, for example, when the manufacturing process became easier and effective pressurised containers were designed later in the century, hydrogen regained wider popularity with the advent of twentieth century airships. This latter situation was brought about mainly by the increasing practice, towards the turn of the century, of diluting coal gas with the 'heavier' carburetted water gas which, while commercially more efficient and cheaper for lighting purposes, had the effect of reducing its load-lifting capability. Furthermore, expectations within the burgeoning aviation industry had changed, bringing the need to lift greater payloads, including motive power units, in addition to people. Until, that is, inflammable gas lived up to its original name once too often!

If early nineteenth-century ballooning itself was essentially a leisurely-paced occupation, then the process of manufacturing hydrogen gas at the point of departure was even more so. It was a lengthy – and dangerous – chemical process requiring the gas emitted from barrels, containing hefty quantities of iron filings onto which copious amounts of sulphuric acid was poured, to be cooled through a large tub of water. Gas thus generated was collected through tubes and piped into the envelope of the balloon through a narrow opening in its neck. Generally fashioned from finest quality oiled silk, this material was made more impervious to the gas by the application of a coating of rubber solution to the inside – yet another French innovation – that, in laboratory tests for gas seepage, was found to be as efficient as the fabric from which modern balloons are made. Conversely, the coated silk was rendered highly inflammable by the turpentine that was used to turn blocks of raw rubber into a liquid.

By such primitive methods the volume of gas necessary to carry one or two people aloft could take up to a day or more to produce and we shall see a little later what was involved in the production of large quantities of hydrogen at the site of a balloon launch. It was, therefore, no frivolous pastime in which these men, rightly described as 'intrepid', were engaged. On the contrary, they were risking their lives at the forefront of discovery in a new branch of science.

Returning now to that day in 1811, the hair-raising events of the next few hours graphically unfold through the words of the voyagers themselves, taken from a contemporary account in *The Gentleman's Magazine*:

> After three minutes of flight the balloon was enveloped by clouds but these were soon cleared. When the aeronauts were at sufficient height they had an extensive view of the surrounding country, with Lichfield, Coventry and Tamworth recognizable beneath them.
> At 2.30pm the barometer [for calculating altitude] stood at 24inches and the thermometer read 50°F. These instruments varied until at 3.15pm they reached their greatest height when the barometer showed 18inches and the thermometer was as low as 38°F.

In those days, before altimeters, a barometer (a mercury-filled glass tube type) would allow a gain in altitude to be measured. A barometer's main use is to measure actual changes in the pressure of air at a particular location, due to the weight of a column of air bearing down upon the column of mercury. As the earth's atmosphere becomes thinner with increasing altitude, the weight of that column of air decreases and the pressure on the mercury will fall. This process allows altitude gained to be deduced by the process of recording the mercury column before take-off then reading it at intervals during the flight, from which the approximate height above ground could be calculated. Where the potential altitude is more than just a few thousand feet above sea level, the calculation of height gained during balloon journeys also involves consideration of the air temperature at the time of each barometer reading. For this flight Sadler seems to have used conversion factors equating to about 1,200ft per inch of mercury fall. In the circumstances of nineteenth-century ballooning this rudimentary process was quite adequate for general information about a flight profile, and its inherent limitations, while obvious now, caused little problem then. The voyage continued:

> In the vicinity of Leicester the wind seemed to have changed to due west and they flew east towards Market Deeping, Lincs, by which point the aeronauts were at their greatest elevation of two and one half miles [13,200ft]. From this vantage point the panorama spread beneath them now included the towns of Peterborough, Stamford, Wisbech, Crowland and Spalding. Mr Sadler, perceiving a current of air passing under the balloon to the northward, deemed it prudent to descend in order to avoid being carried towards the sea, now visible in the distance [possibly a view of The Wash]. Having become quite distended, it also became necessary to let out some of the gas from the balloon, which was done at intervals, until it descended into the air current Sadler had previously noticed. The travellers were then carried in a northerly direction and with Spalding now on their right and Bourne to the left, Sadler stabilized the craft by throwing out all the remaining ballast. This, however, failed to arrest the balloon's descent and the carriage struck Mother Earth at the hamlet of Burton Pedwardine, near Heckington, south of Sleaford. Grappling irons on the end of a rope were thrown out but these, too, proved ineffective.
>
> On the second impact with the ground Mr Sadler, having hold of the valve-line was, by a sudden jerk caused by a grapple holding momentarily, thrown out violently from the carriage. He received several contusions about the head and body but had sufficient presence of mind to call out to Mr Burcham not to quit his seat! Being lightened by the sudden removal of Sadler's weight, the balloon swiftly rose to a height of 300 feet, much to the alarm and great hazard of the poor gentleman left on board. At length he succeeded in gaining a hold of the valve-line, which he pulled hard, thereby releasing sufficient gas to descend once more. Throwing out a grappling hook, he was able to prevent the balloon from rising but its erratic progress was finally brought to an abrupt halt when the, by now, sagging envelope wrapped around a tree and was flogged to shreds in the strong wind.

Mr Burcham was fortunately relieved from his perilous position, being deposited on terra firma safely and with only a slight bruise to mark his passage. He had 'landed' in the parish of Asgarby, two miles from where he and Sadler parted company.

Quickly surrounded by people anxious to render assistance Burcham was speedily conveyed to the home of Mr Godson, upon whose acres he had landed, where he was 'welcomed heartily and with good cheer'. From there he proceeded to Heckington where he and Sadler were reunited in the market square. The interview may be more easily pictured than described, each believing the other had been killed, 'that their raptures on meeting were almost ludicrous'. Sadler himself had, on recovering from his fall, proceeded to the nearby house of a miller before making his way into Heckington. Duly recovered from their exploits, the two aeronauts made a more leisurely departure for Birmingham in a chaise and four at 3 a.m. the next morning.

Landing was always the most dangerous part of an aerial journey. Sadler had learned this the hard way. Back in May 1785, during an ascent from Manchester, he flew fifty miles to Pontefract (Yorkshire) only to be thrown out of the car when it hit the ground violently.

Relieved of his weight naturally the balloon surged back into the air and was whisked away on the wind until it finally flopped to earth again miles away in Gainsborough (Lincolnshire). Sadler had several narrow escapes in 1784 and 1785, which was another factor contributing to his long break from ballooning shortly after. His passion for more 'earth-bound' invention occupied him quite safely and for many years he devoted much time and energy to improving the steam engine, coming up with designs for a rotary steam engine and steam-powered road vehicles, until after a break of twenty-five years he was encouraged to take to the sky once again.

In the course of their epic journey of 1811, Sadler and Burcham calculated they had flown a distance of 112 miles from Birmingham in one hour and twenty minutes. This splendid achievement was due, of course, to exceptional wind conditions thrusting them along at the remarkable average ground speed of 84mph. Even so, it took great nerve and daring to ride that wind for such a distance and height, and the speed was not matched for another century. It is a fitting tribute to the bravery of the balloon pioneers, yet it will be seen that those men and women who followed Sadler into the skies displayed, in their own way, no less daring than he.

James Sadler made only one subsequent aerial visit to the region, to Deken's Lodge, Pickworth, near Stamford, on 1 November 1813, which was the termination point for the first manned flight made from Nottingham.

James Sadler making the first flight from Nottingham on 1 November 1813. (Spalding Gentlemen's Society)

Beneath the intriguing headline 'The Balloon and The Chase', *Stamford Mercury* reported that 'lovers of hunting who had the good fortune to be out with Lord Lonsdale's hounds that day were treated to a new and novel species of "game", namely a balloon'. Having had what he described in a typical understatement as an 'uneventful' thirty-four-mile journey across the vale of Belvoir from Nottingham, where he had ascended from a canal barge moored at the Navigation Company's wharf, Sadler came down near Mr William Deken's house, slap bang in the middle of the Cottesmore Hunt in full cry!

Reaching an altitude of 18,500ft during his one-and-a-quarter-hour flight, the poor fellow was shaking and numb with cold when he stepped out of the car. He had passed through a dense rain cloud and icicles had first formed on his cloak then melted, soaking both his clothes and the floor of the car as he descended. In the true style of gentlemen of the day, he was conveyed to the warmth of Mr Deken's home, where with grand hospitality thrust upon him he quickly recovered his composure and regaled his hosts with the story of his experience.

What set Sadler apart from his contemporaries and rivals in Britain and France, is that he was acknowledged – if a little generously – as the first to consciously apply some scientific principles to the new activity of aerostation.

Following his abortive attempt to fly across the Irish Sea, the *Dublin Magazine* of November 1812 paid the following tribute to James Sadler:

> Whoever peruses theory of winds must be convinced of the applicability...of air currents to the guidance of balloons. This appears to have been the grand object of Mr Sadler in his ascents. Montgolfier, Lunardi and almost the whole tribe of balloonists seem to have been totally unacquainted with this. Many of them performed their voyages, ascended and descended, raised the stare of astonishment in the vulgar, set in motion the tongues of the titled and fashionable non-entities of the day, but made not (because they were incapable) one single observation that could promote science, or enlarge the boundaries of human knowledge. Mr Sadler (if report speak truth of him), either knew when he first ascended or has since been taught, somewhat of the rudiments of the sciences connected with aerostation. We may, therefore, look forward to his future ascents for the much-wished-for discovery of some method to render the balloon of important national service. In the event of sieges and battles, indeed in almost every military purpose, it is impossible fully to estimate the mighty advantages that would arise from gaining knowledge of its complete management. Discovery may extend itself at every point and more may be gained by a future age in the progress of knowledge than can possibly enter the calculations of the present.

Sadler, although keen to identify, utilise and chart the existence of air currents, had to be content to place his trust in the wind and fly wherever it chose to carry him. James made his last ascent in September 1816 and died peacefully in Oxford on 26 March 1828.

In the wake of euphoria created by the likes of the Montgolfier brothers, Vincenti Lunardi and Jacques Charles, the year 1785 witnessed a veritable flock of aeronauts taking to the sky over Great Britain. These included Sadler, of course, from, to list a few examples, Oxford, Manchester, Worcester and Lichfield; Lunardi from London, Liverpool and Edinburgh; Zambeccari from London; Harper from Birmingham; Deeker

James Sadler by an unknown artist, *c.* 1820. (National Portrait Gallery, London)

and Money, both from Norwich; Crosbie from Dublin; Cracknell from Nottingham, and Baldwin from Chester. Let us not forget, too, the first aerial crossing from Dover to Calais on 7 January 1785 by Jean-Pierre Blanchard and Dr John Jeffries. Yet after this first flush of enthusiasm in England and France, so prolific was the activity in the space of less than three years, that it became quite commonplace and public interest in the discovery of flight actually became apathetic – a phenomenon that can be detected at several points in its history. Furthermore, when possibly the most famous man to fly up to that point – Pilatre de Rozier – and his companion died in a ballooning accident on 15 June 1785, there was a public outcry and many said it was retribution for trying to do what man was not ordained to do. These factors, coupled with the upheaval of the French Revolution and then a long period of war between England and France, served to put public and private aeronautical activity on hold in those two countries for nearly twenty-five years, except for a flurry around the time of the 'phoney peace' brought about by the Treaty of Amiens. The resurgence of Sadler's flying career from 1810, therefore, can reasonably be regarded as signalling the beginning of the second phase of the story of ballooning.

Coal, that black gold and fuel of the Industrial Revolution, helped usher in the factory age of the early 1800s. In addition to providing the basis of steam propulsion for railways, ships and machinery, its major by-product – coal gas – made it possible to illuminate homes and work places in the burgeoning towns and cities of England, and consequently new gas works companies proliferated. In 1821, the name of Charles Green rose to prominence in nineteenth-century balloon aeronautics and remained at its forefront for the next thirty years. Charles Green was instrumental in discovering and exploiting the ability of coal gas to be used as a lifting agent, first employing the gas in a balloon in that coronation year.

It will be seen that this eminent professional aeronaut well and truly imprinted the new science of ballooning in the minds of the people throughout the land and in doing so endeared himself in their hearts. Indeed, his enormous contribution to the public awareness of aeronautics would be equalled only by exploits of the likes of B.C. Hucks and Alan Cobham a hundred years later. Some historians have suggested that 'he was uneducated' and 'he was not a scientific aeronaut', but while it is true that he was not and would never claim to be a 'scientist' in the mould of Cavendish, Black or Glaisher, he certainly deserves great credit for discovering what it took to be able to venture into the air efficiently and return to earth safely, time after time, in all weather conditions. Green also introduced a number of technical innovations that to some were merely 'practical' rather than 'scientific', but that could – in the age of technical discovery that was the nineteenth century – be regarded as a matter of semantics. We shall see from the examples of his exploits that follow, his vast experience – freely and widely reported upon during his lifetime – undoubtedly made the 'scientific' claims of those who followed (Coxwell with Glaisher in the 1860s, for example) a great deal easier to achieve. As Shakespeare said, 'some are born great; some achieve greatness and others have greatness thrust upon them'. Charles Green certainly achieved greatness through his own efforts. All this naturally gives rise to the question, how did Green arrive at his particular technique for ballooning and his unique contribution to the world of aeronautics?

Charles Green was born the son of a London fruiterer, on 31 January 1785 in Goswell Road, north London. It was not until 19 July 1821 that he made his first balloon ascent, as part of the celebrations marking the coronation of King George IV in London, displaying a high degree of confidence in his own ability that would become his trademark. Furthermore, Green chose this as the occasion to make the first public use of coal gas to inflate a balloon, when hitherto hydrogen or heated air had generally been used for

Charles Green (1785–1870) painted by Hilaire Ledru in 1835. (National Portrait Gallery, London)

that purpose. The balloon was inflated from the gas main running down Piccadilly and its inaugural journey began in Green Park and ended just north of Barnet. His earliest advertised ascent had in fact been arranged for the previous day, from the Belvedere Gardens, Pentonville, but he cancelled this in order to accommodate the much more prestigious royal assignment.

Green was a practical fellow with an enquiring mind who went into his father's business after leaving school. His decision to use coal gas (also known as carburetted hydrogen gas) in a balloon is reputed to be one of the outcomes of his experiments to make coal gas with which to illuminate their business premises in Clerkenwell Road. Coal gas is a by-product of the destructive distillation of soft coal during the process of making coke. Its basic composition is largely hydrogen and methane with additional small quantities of other hydrocarbons, carbon monoxide (the poisonous substance), carbon dioxide and nitrogen. He found that gas produced during the initial stage of the distillation process gave off the brightest light when ignited but gas produced towards the end of the distillation cycle gave off a barely visible flame. Green believed this anomaly indicated that the later-stage gas contained a high proportion of hydrogen. He would have been aware of the value of hydrogen to aeronauts, not least because he lived and worked in the vicinity of the City Road, a short-distance from where many famous aeronauts of the day made ascents. For example, the Eagle Tavern garden in City Road (the very same 'Eagle' in the nursery rhyme 'Pop goes the weasel') was the venue for James Sadler in 1810 and 1811, and it continued to be a popular ballooning venue during Green's heyday. Similarly, Lunardi and subsequent aeronauts used the artillery ground in Moorfields, also quite close to Clerkenwell.

Charles Green's innovation produced a most significant change to the science of aerostation and illustrates his foresight in utilising a gas that, by 1817, could be manufactured easily, cheaply, in bulk, and to a reasonable standard of consistency and eventually would become universally available. A genial fellow on the ground and a 'tartar' in the air, Green is widely acknowledged to have made over 500 balloon ascents between 1821 and 1852. Indeed, the pursuit appears to have had a special attraction for the whole Green family. It is claimed that no less than ten persons of that name made balloon ascents in the early nineteenth century, including his wife Martha, brothers Henry, James and William, Charles's son George, who himself made eighty-three aerial journeys, and another relative named Robert.

The coronation flight was a great success but even this landing was fraught with danger when he was thrown out of the car (sounds familiar!) and dragged along the ground for a quarter of a mile while clinging to the netting hoop. Like any pilot's first few sorties, if you survived the first few, the knowledge gained improved one's survival prospects. It took a while to get the hang of making a 'soft' landing and Green and his contemporaries were no exception. On his third flight, on 1 October 1821, he narrowly escaped drowning when his balloon came down four miles off Brighton and he was fished out of the sea by the packet ship *Thomas*. Only a couple of his first seven flights were without landing mishaps and indeed one put him out of action from July 1822 to June 1823. Even though he learned to master his landings it was a while before the public learned to treat coal gas with a measure of respect, particularly in the quantities being vented off on landing! Such perils were amply demonstrated at Romford on 3 June 1823, at the end of Green's first trip after his layoff. The *Stamford Mercury* reported:

> Mr Green ascended with his balloon at 7.45pm from the gardens of the Mermaid Tavern, Hackney, London and landed safely at 8.20pm four miles north of Romford, Essex. The evening was fine and the balloon was in sight of the metropolis until it landed. It rose to a height of two and one half miles [13,200ft] where the cold was so intense that Mr Green's

fingers were paralysed till he could hardly pull the gas release valve. The farmer into whose field of clover he descended detained him until he paid for damage to the crop, which was destroyed by the crowds of country folk who flocked to see and assist him.

A circumstance occurred which nearly proved fatal to the lives of two people and shows how pernicious is the coal or carburetted hydrogen gas. Mr Green was sitting in the car and the balloon was discharging its contents very fast through the valve which was wide open. Two men came up to render assistance and one took hold of the valve and breathed deeply of the gas being exhausted. He inhaled so much of the noxious gas that he fell insensible to the ground. The condition of the man was not observed by Mr Green and another man came forward to render assistance to the aeronaut and met with the same misfortune and fell to the ground apparently dead. By this time the balloon was nearly empty and the men were seen and carried away to some distance where they were resuscitated with difficulty.

He learned from his mistakes and his ability along with his personal reputation grew rapidly thereafter.

Charles Green's first venture into the East Midlands region is recorded as 8 July 1824 and was his sixteenth flight. Ascending at 5 p.m. from the gas works in Northampton, in a balloon 107ft in circumference and 38ft in height, containing 136,000 gallons – equal to 22,500cu.ft – of coal gas, he landed at Soham, near Ely, two and a half hours later. This craft was the balloon made from alternate gores of blue, crimson and gold silk emblazoned with the name *Royal Coronation*, as used on that earlier royal occasion and whose image featured in Green's newspaper advertisements.

Reporting this new phenomenon, the *Stamford Mercury* newspaper suggested 'much consternation was caused at Soham by the extraordinary appearance that evening of something in the air, nearly if not a mile from the earth.' It was at first sight imagined to be an immense kite but it was afterwards considered to be a fancy balloon, sent off from Cambridge. Both of these conjectures were, however, quite wrong. The 'thing' proved to be a balloon with a car containing the celebrated aeronaut himself.

Green intended to land near Haddenham, near Ely, and opened the gas valve to begin his descent as the village hove into view. However, as he went lower a great deal of water could be seen below in the area known as Grunty Fen, so he threw out some ballast to avoid a 'splash-down'. Having then passed over Ely and Soham itself, he was able to alight safely two miles from the village. His landing was not without difficulty and a little danger from a quite unexpected quarter.

At first the grappling iron he threw out would not hold in the light fenland soil. As the rope trailed over the ground Green espied a labourer close by, working on top of a haystack. Calling out to this fellow to lend a hand, the man was so startled by the stealthy arrival of this apparition that he threatened to run his pitchfork through both Green and his balloon if he dared approach closer! Throwing out the grapple once more, it fastened onto a gate and held, thus allowing the balloon to sink safely to the ground, where he was this time assisted by more friendly natives making hay – no doubt while the sun still shone! Far from being upset by his brush with the aggressive labourer, Green's verdict on the journey was:

The day being perfectly serene added very materially to the grandeur and magnificence of the scene. My extreme height was two and one half miles [13,000ft] and the distance covered was fifty miles in the 2½ hours since my ascent began.

Deflation complete, the balloon was placed in a cart and the party retired to the Crown Inn, Soham, for refreshment. Suitably rested for a few hours, Green caught the post chaise to Northampton, arriving at 1 p.m. the next day, Friday, to be greeted by the ringing of church bells and the most cordial civic congratulations.

Stamford, situated at the western edge of the Fens, became a popular venue for ballooning during the nineteenth century and Saturday 2 July 1825 in the town dawned a glorious mid-summer day with not a cloud in the sky. Charles Green and his balloon were objects of enthralled attention from quite the largest crowd ever assembled in that town. An estimated 25,000–30,000 people, many coming from as far as fifty miles away, assembled to witness this, the first balloon ascent from the town. Inns were filled to overflowing, few private houses were without guests and a carnival atmosphere pervaded the whole proceedings.

After a few years as a professional aeronaut Green had, by now, developed an astute commercial approach to his exhibitions. The *Lincoln, Stamford and Rutland Mercury*, that oldest and most respected of provincial newspapers, carried advance notice of the event and in view of its vast catchment area, made a practice of reporting on all Green's (and others') ballooning activity making it a rich source of detailed information on nineteenth-century aeronautical activity.

Focus of attention was on the gas works enclosure where admission on the day was 2*s* (10p). Stored in the Assembly Rooms for three days prior to the launch, partially inflated with atmospheric air, the balloon was put on show to the public. An additional attraction took the form of a gas-filled model balloon, made from gold-beaters skin which was made to ascend and descend within the confines of the Assembly Rooms. Tickets to enter the rooms were on sale, 'at one shilling [5p] for Ladies and Gentlemen and sixpence [2½p] to Children and Servants', and those prices would represent a considerable sum to the ordinary person in those days. Nevertheless, £100 was taken during the three days and a further £200 at the gate on the day of the ascent but for the majority, peering from the streets or house windows and roofs, it was a free show. However, it was reported that Mr Green and his companion for the trip, Miss Stocks, would be left with a handsome remuneration for their hazard.

The origin of Miss Stocks is not commented upon in the press but it is most likely the same fair lady who, little more than a year earlier at the tender age of eighteen, had undergone a terrifying ordeal of crashing to earth in a balloon in which she was a passenger. Her companion on that occasion, Thomas Harris, was killed in the accident. If it was the same Miss Stocks then she certainly had pluck to go up again, although she could not have been in safer hands.

On the question of remuneration, some indication of the vast sums of money such men could command can be gained from reports that in 1784 Vincent Lunardi received a £2,000 profit share after making the first flight in Britain; the entrance fee to his take-off ground was no less than 1 guinea (£1.05). In 1785 Dr John Jeffries not only invested money towards the capital cost of that first aerial Channel crossing project but was also charged a fee of £100 for the privilege of accompanying Blanchard. James Sadler, in 1811, offered to carry a passenger aloft for the sum of 100 guineas (£105) – and indeed a certain Captain Paget RN took up that offer. Green himself charged a passenger, Isaac Sparrow, £50 in Oxford in 1823. Costed at equivalent modern rates these represent phenomenal sums of money.

By 3.30 p.m., inflation was complete and Charles Green, accompanied by the doughty Miss Stocks, stepped aboard, cast off, and balloon and passengers rose into a cloudless sky amid tumultuous applause. They responded graciously by waving brightly coloured flags. With a gentle wind from the west the balloon rose quietly and steadily, remaining in sight of the watchers below for an hour. For a graphic description of Mr Green's journey we are indebted once again to the *Stamford Mercury*, who quoted him thus:

My course was east by south at first. Being fully inflated, some gas was displaced from the balloon as we rose higher and thus the lifting power was decreased. This was compensated for by throwing out a quantity of ballast and eventually an altitude of 7,500 feet was reached and maintained as the balloon passed over the villages of Ufford and Glinton.

A most splendid view of Peterborough cathedral was afforded us some five miles away on our right hand and in a short while we passed over the south side of Thorney. I decided that, as the surrounding countryside was favourable for a safe descent and numerous people could be seen running towards us, we should descend at this place. No time was lost, therefore, in making a controlled descent. With little forward motion, the carriage came to earth in a field of standing corn. As I did not wish to damage the crop I decided to allow Miss Stocks to alight with the flags and recording instruments, before proceeding to another, clearer, field in order to deflate and dismantle the balloon. Those good people who had assisted my descent by holding on to the trail ropes were urged to release their grip, whereupon I re-ascended. Unbeknown to me, one of their number, a labourer by the name of Bolton, taken by surprise was carried aloft still clinging to a rope. He retained the presence of mind to hold his grip but fortuitously it required only a passage of about a mile before I was able to descend in to a grass field belonging to George Maxwell, Esquire. As the balloon sank gently towards the ground Mr Bolton judged his moment and dropped to earth from a height of fifteen feet. He was stunned only momentarily and this splendid fellow recovered quickly enough to follow me once more in order to render further assistance to secure my balloon.

Miss Stocks joined me within minutes of my arrival and we were received most hospitably by Mr Crane whose residence was close by. A chaise and four was generously procured for us by our host and we returned to Stamford by 9.30pm.

This relatively short but eventful journey of some fifteen miles in all marked the end of Charles Green's thirty-fifth flight and the beginning of his lively aerial association with the region, which was to continue for the next decade.

Naturally, Green's 'handsome remuneration' was dependent to a large extent upon the price of gas charged by the local, privately owned, gas company and his 'gate' receipts. Indications of coal gas prices across the East Midlands have been found as follows:

Stamford	1826	10/6	(52½p) per 1,000cu.ft
Boston	1826	12/6	(62½p)
Stamford	1828	11/-	(55p)
Wisbech	1832	15/-	(75p)
Leicester	1837	7/6	(37½p)
Norwich	1837	10/-	(50p)
Wisbech	1838	7/6	(37½p)
Norwich	1840	7/-	(35p)
Boston	1855	4/-	(20p)

At this time the well-known husband and wife aeronautical duo, George and Margaret Graham, were among those still persisting with hydrogen-filled balloons. In August 1825 they happened to be visiting Norwich and though their aerial destinations on that occasion would normally not qualify for inclusion in this story, in this instance their 'antics' provide a convenient and highly enlightening opportunity to compare financial, practical and presentational differences between Charles Green's approach and that of one of his 'rivals'. Both parties were at a similar point in their careers as professional aeronauts, in that Green's trip from Stamford was his thirty-fifth while that of Graham from Norwich was his thirtieth.

In financial terms Green's event occupied him in Stamford for four days during which his income was £300. His coal gas would have cost him about £13 and of course he would have had other expenses such as the hire of the hall, seating, band, accommodation, travelling and so on, but he would have cleared well over £200. George Graham and his wife, on the other hand, had to work hard for about a month and were unlikely to have made that amount of profit, if indeed they made any at all.

The Grahams' main problem lay in the poor quality of hydrogen produced by George's inefficient production system. This resulted in weak lifting power, delayed schedules and problems coinciding with bad weather.

Bad weather and weak gas dogged George Graham's scheduled ascent from Richmond Hill Gardens in Norwich on Saturday 6 August 1825. Heavy rain, high winds and thunderstorms kept spectators away, and only 300–400 souls paid an average of 3s (15p) each to go into the gardens. Already contemplating a hefty loss, Mr Graham set off to satisfy honour all round but landed just thirteen miles away in Beighton. Evidence of his financial loss appeared in the newspaper when he was encouraged by the local organising committee to appeal to the conscience of the public and arrange another ascent to try to recoup his losses. The press advertisement for a further ascent appeared in the form of a table of disbursements showing outgoings of £203 and receipts of just £93, leaving his net loss as a thumping £110.

His second attempt was scheduled for Monday 15 August, with entrance prices of 2s 6d (12½p) and 1s 6d (7½p). It was a fiasco, with the balloon unable to rise due to poor gas quality and wasted gas through an inefficient inflation system. The Grahams were becoming something of an embarrassment at this point when, fortunately, they had to leave Norwich to fulfil a prior engagement in Chelmsford on Thursday 18 August. This event, too, was a failure due once again to the poor lifting quality of the hydrogen gas.

32d ASCENT.

MR. GRAHAM,

BY the advice of his Friends, respectfully informs the Nobility and Gentry of Norwich, and the Public in general, that he purposes to make his SECOND ASCENT from RICHMOND HILL GARDENS, on MONDAY next, the 15th of August, at Four o'clock in the Afternoon. Admission to the Inflation and Ascent, 2s. 6d.—Children 1s. 6d. each.

Mr. GRAHAM,

Having lost a considerable Sum of Money on Saturday last, in consequence of the unpropitious state of the weather, is induced by the advice of several most respectable Gentlemen of the City and County, to lay the particulars of his Receipts and Disbursements before the Public, having been assured by those Friends, that this Appeal will prevent his leaving this City with so severe a loss.

☞ *Subscriptions will be received at the Printing Offices.*

DISBURSEMENTS.	£.	s.	d.	£.	s.	d.
Vitriolic Acid and Carriage	57	15	11			
Iron	13	0	0			
Carriage of Iron	8	9	2			
Carriage of Balloon, &c	6	11	8			
Tank, Stage, and Labour	30	0	0			
Five Rum Puncheons	4	0	0			
Two ditto	2	0	0			
Printing, Advertising, &c	10	0	0			
Delivering Bills	1	5	0			
Music	6	0	0			
Officers	3	14	0	263	12	5
Expences on Descending & Returning to Norwich with balloon	4	5	0			
Mr. and Mrs. Graham's Journey to Norwich	3	8	0			
Mr. Lane, three week's wages	6	6	0			
Water and Carriage	1	11	6			
Attendance at St. Andrew's Hall	3	0	0			
Mr. Lovick's Bill	6	7	7			
Journey from and returning to London Eight Assistants	18	17	9			
Mr. Nobbs's Bill	9	5	8			
Mr. Lane's Bill, for Labour, &c.	7	15	2			

RECEIPTS.	£.	s.	d.	£.	s.	d.
St. Andrew's Hall	23	7	6			
Day of Ascent in Gold	13	0	0	92	17	0
Do. in Notes	19	0	0			
Do. in Silver	37	9	6			
Loss				110	15	5

The Vouchers are deposited in the hands of the Printers of this Paper, for the inspection of the Public.

BY PERMISSION,
And under the kind Patronage of the Right Worshipful the MAYOR.

NOTICE,

THE FINAL AND POSITIVE CLOSE OF THE

EXHIBITION AND PROMENADE

Now Exhibiting with the most unbounded and general approbation in the Assembly Rooms, Norwich, will be on SATURDAY, August 20th, *after which it will be Removed.*

MADAME TUSSAUD,

ARTISTE,

RESPECTFULLY notifies, that after the above period her Collection will be Removed, she states this more particularly in order that no one may be disappointed in viewing an Exhibition allowed to have no equal in Europe.

M. T. would be deficient in gratitude were she not to return her most sincere thanks to the Right

Advertisement for 15 August and profit and loss account relating to various attempts by Mr and Mrs Graham to organise a balloon display in Norwich in August 1825. (*Norwich & Norfolk Chronicle*)

The ascent was postponed to the next day but even then it turned into a farce for similar reasons. His hydrogen gas was just not good enough. George Graham and his companion for the journey, Captain Gape of the Scots Greys, sat in the car but the balloon refused to budge. After heated discussion it was agreed that Margaret Graham would take the place of her husband in the hope that her lesser weight would do the trick and she was actually no stranger to piloting a balloon on her own account.

Still the balloon did not rise and this was met with yet more heated discussion because the crowd did not wish to see Mrs Graham ascend with a complete novice. The cry went up for George Graham to make the ascent alone, which was tried – without success. Graham climbed out of the car and now the crowd roared for Captain Gape, a slightly built man, to try his hand. Bravado got the better of him and he jumped into the car, cast off the rope and – lo! – the balloon rose crabwise into the air, snagging on a chimney of the House of Correction as it did so. Captain Gape managed to clamber out of the car onto a gable end and, freed of all weight, the balloon rose crewless into the sky. An hour later it flopped to earth near the Rodney Inn at Little Baddow in Essex from whence it was returned, slightly torn, to the hapless Grahams.

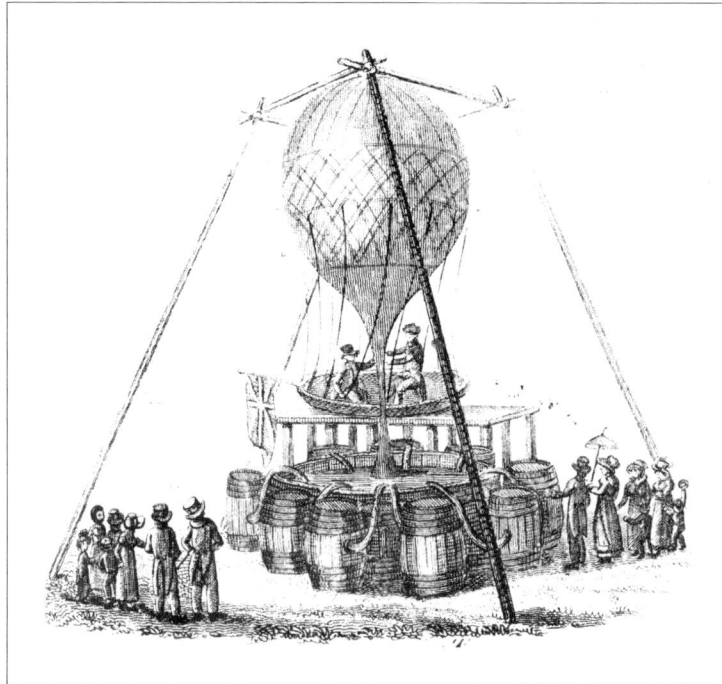

Manufacturing hydrogen gas at the site of an early nineteenth-century balloon event. The barrels contain iron and sulphuric acid and the resultant hydrogen gas is piped through a tub of water to cool it before entering the neck of the balloon.

The saga was not over yet. Graham was committed to return to Norwich for another attempt to recoup his losses and, indeed, try to recover his reputation. Raising the viewing price to 3s and 2s on Tuesday 30 August helped a little but the crowds came more to see what sort of a mess the aeronauts would make this time. Colonel John Harvey, the high sheriff of Norfolk, was to accompany George Graham, but he had little chance because the same old problems reared their head again. So determined was Colonel Harvey to get into the sky he loudly proclaimed he would go up alone. The cry came back, 'No! He shall not go up alone!' Captain Gape was in attendance and, no doubt emboldened by his escapade at Chelmsford, he leaped into the car with John Harvey and there they both sat waiting for? Nothing!

Fed up with waiting they both disembarked and George found there was just enough lift to allow him to ascend on his own and travel one measly mile to Earlham on the outskirts of the city. The good colonel's ego was sadly dented by these fiascos and as the leader of the organising committee, he made it clear that the Grahams would make a good ascent – with him as a passenger – come what may. It had become a matter of personal honour! To this end Colonel Harvey drafted in his own 'eminent' chemist to oversee production of the hydrogen gas under the strictest controls; it was to be ready to take to the air on Tuesday 6 September – without fail! Eminent chemist or not, the balloon still would not lift George and the colonel into the sky.

It fell to Margaret Graham to save face for all concerned. A sunny day brought a packed crowd to Richmond Gardens and the gas was good for a change. Colonel Harvey was in his element and he achieved his ambition to place himself where he felt he rightfully belonged – way above the heads of all others! The pair flew as far as Brundall, covering the eight miles in twenty-five minutes without any further mishap to spoil Colonel Harvey's euphoric state.

There was no further public assessment of the financial outcome of the Graham's month in East Anglia, but they probably made good their early losses. As for a comparison between Green and Graham – well, that speaks for itself.

Mr Green begs leave most respectfully to announce to the gentry and inhabitants of BOSTON and its vicinity that he intends making his fifty-second ascent, from the Gas Works, Boston on Thursday June 8 1826, at 3 o'clock in the afternoon.

Thus ran the balloon-emblazoned advertisement in the *Stamford Mercury* of 2 June, signalling Charles Green's return to the region.

Once again the partially inflated balloon together with its 'car and appendages' was on view to the paying public the day prior to the event. Tickets to witness the process of inflation and launching of the craft could be had for the sum of 2*s* 6*d* (12½p). Ladies and gentlemen were assured that secure seats were to be erected for those honouring Mr Green with their presence in the gas works enclosure, where a band of music would also entertain them. 600 local gentry were induced to part with this not inconsiderable entrance fee and availed themselves of the spectacle at close quarters. Meanwhile a crowd, estimated at over 20,000 people, filled the streets of Boston to overflowing to catch a glimpse of this wonder of the age. On this occasion Mr Green chose to be unaccompanied during his ascent, although no reason was given for his decision.

It is not often that a first-hand account, other than that provided by a principal or journalist, can be found of such an event in the early nineteenth century. John Peck, gentleman, farmer and patron of all things modern, of Parson Drove, near Wisbech, kept a detailed diary every year from 1814 until his death in 1851. The following extract is one of several references to ballooning found in his diaries, now in the custody of Wisbech Museum. Of Green's visit to Boston, John Peck, in his beautiful copperplate style, wrote:

8 June 1826. Started at 6[a.m.]. Breakfasted at Kirton. On with Mrs Peck and a party to Boston to see Mr Green ascend with his Balloon, which was indeed a truly Grand sight.

I offered him £5 to go up with him but he would take no one.

Never saw Boston so full of Company. Returned and tea with Mr Palethorpe. Home to Parson Drove by 11 o'clock [p.m.].

The sum of £5 was a prodigious inducement in those days but Green was not a man to let greed override his better judgement.

On first leaving the ground the balloon took a south-westerly course which in a short while changed to due west. This course carried Green between the villages of Swineshead and Heckington, in a direct line for Grantham in the vicinity of Belton Park, the home at that time of Sir John Thorold. Green commenced his descent while still to the east of Grantham. Having passed the town, on reaching 4,000ft he encountered a south-easterly air current which now swung him north in the direction of Newark. However, as the countryside was still favourable for a landing, he continued to descend coming to rest at

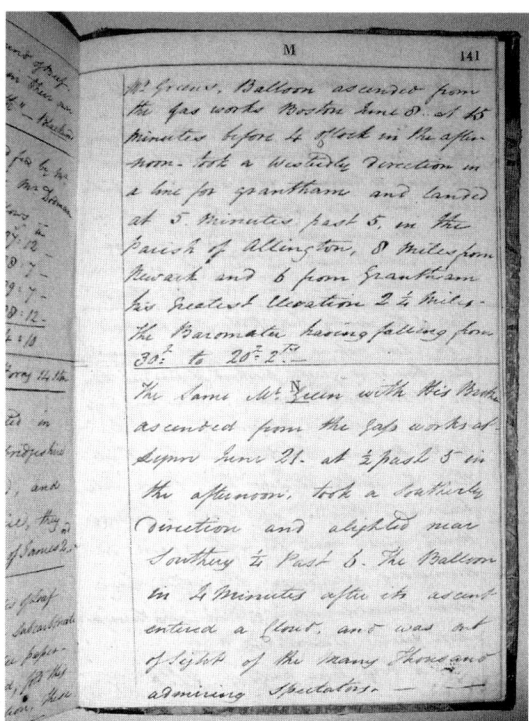

A page from John Peck's diary in which he recorded meeting Charles Green in Boston and seeing his balloon in flight on several occasions near Wisbech. (Wisbech Museum)

5.05 p.m., in the grounds of the home of T. Earle Welby Esquire, in the parish of Allington near Bottesford.

Green had, on this occasion, travelled thirty miles in two hours, and his barometer readings showed a drop in pressure from 30in on take-off to a low of 20.2in, indicating a gain of about 12,000ft at his greatest elevation. By the time he had dismantled the balloon and car several gentlemen arrived on horseback from Grantham to acquaint themselves with this celebrity. One of them kindly lent Green his mount, while another horse belonging to Mr J.B. Tunnard of the Blue Lion Inn, Grantham, had the balloon packed upon it and conveyed to the George Inn, where Green himself was greeted warmly by a large gathering of local gentlemen. Mr Welby sent an invitation to the aeronaut, which he accepted and 'partook of a sumptuous dinner' that same evening. Charles Green, duly replete and basking in the adulation of the gentry of Grantham, transferred his trusty balloon once more into a chaise and four, departed Grantham at 10 p.m. and arrived back at the Peacock Inn, Boston, at 3 a.m. next morning, where no doubt he enjoyed a much deserved sleep.

Although capitalising on his fame and touring the length and breadth of the country making ascents, the year 1826 saw Charles Green spending much of his time in the Fenland region, probably due to the major towns, even in this predominantly rural area, vying with each other in an attempt not to be left behind or considered unprogressive. The spectacle of ballooning fed by a local gas supply presented ideal opportunities for town councils to draw attention to their civic efforts. It was this very context that brought Charles Green to King's Lynn for the first time on 21 June 1826, for the opening of the town gas works and his fifty-fourth aerial journey.

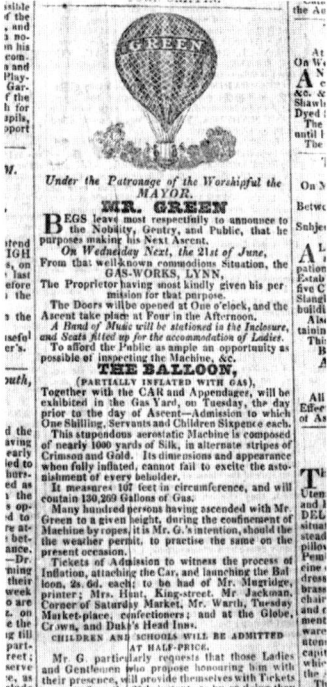

Advertisement for Charles Green's fifty-fourth flight, to take place in King's Lynn on 21 June 1826. (*Norwich & Norfolk Chronicle*)

All roads to the town were jammed with people on foot and horseback, as well as in carriages of every description. Accompanied on this auspicious occasion by his brother George, Charles Green's reputation by now could guarantee a large audience anywhere in the land.

It was 5.30 p.m. when the aeronauts rose majestically into the air beneath their magnificent balloon in the presence of an estimated 15,000 crowd. Drifting southwards, gaining altitude, they were carried towards Downham Market. After a flight lasting an hour the Green brothers descended without mishap on the estate of Robert Martin Esquire, in the parish of Southery, six miles from Downham. The party was taken to that town where, upon their arrival, church bells rang out in their honour. Receiving the congratulations of a group of local gentlemen, lead by C. Hunter Esq., by whom they were most hospitably entertained, Charles and George returned to King's Lynn in a chaise and four where they arrived tired but elated at 9.30 p.m.

John Peck did not visit King's Lynn on this occasion but expressed a keen interest in the event described above. His diary for that day reads:

21 June 1826, Rool [sic] and harrowing, 11 acres sown. A cold day. Mr Green ascended from Lynn with his Balloon and alighted near to Southery. I went up Parson Drove church with a telescope expecting the wind to drive him this way but turning more to the north, was disappointed.

Louth was the next eastern town to enjoy a Charles Green exhibition. At 3.45 p.m. on Friday 4 August 1826, he lifted off – solo – from the Wool Mart Yard and, climbing steadily, the balloon drifted off to the south-east. Following the edge of the Lincolnshire Wolds, Green passed over Alford where, having reached 12,000ft in altitude with the temperature at twelve degrees of frost, upper level winds began to carry the balloon towards the sea. Green lost altitude to regain the more favourable lower level breeze but severe up-currents at this point made his descent a rough and very tiring affair. Lower down he flew towards Spilsby and made a good landing in Mr Goodwin's pasture one and a half miles north of the town. As usual many people followed his progress and helped gather up the balloon. The Reverend John Banks pressed Mr Green to take refreshment with him in Spilsby before the aeronaut proceeded to Harrington Hall, the mansion of R. Cracroft Esq., to receive the adulation of the family and a party of ladies who had witnessed the ascent. After what was almost the inevitable sumptuous meal he returned to Louth at 2a.m. next morning to yet another rapturous welcome, this time by the members of the organising committee at the New King's Head Inn. What a grand life!

From the voluble wording used in his next advertisement, an explanation for Green's solo ascent at Boston in June 1826 can perhaps be deduced. The cumbersome text read:

> In consequence of numerous solicitations from several respectable individuals of Boston and from the circumstance of hundreds being disappointed of witnessing his last ascent, owing to the injury which his balloon sustained obliging him to ascend earlier than he wished, Mr Charles Green most respectfully announces to the Gentry and inhabitants of Boston ... that he intends making his fifty-ninth ascent with his new, beautiful and stupendous balloon Royal Coronation on Wednesday August 30 1826.

In what would today be described as 'hype', it was also announced that he would be accompanied by an unnamed lady and gentleman, well-known in Boston and its neighbourhood. Furthermore, should the day's weather prove calm, a parachute containing a live animal would be dropped from the car, to descend in safety into the enclosure. Topping this, and weather permitting, the townsfolk of Boston could emulate the worthy citizens of places as York, Worcester, Warwick, Newcastle and Shrewsbury, by ascending with Mr. Green who would be offering tethered flights. This was indeed bringing aeronautics to the masses!

When the great day dawned, however, the wind was considered far too strong and this prevented both the tethered flights and the parachute drop. The unnamed lady who was to have ascended with Mr Green had second thoughts and declined the trip but eventually Mr Henry Brooke Jr, editor of the *Boston Gazette*, accompanied him aloft. Their journey was uneventful and one hour and ten minutes later, at 5.15 p.m., a safe landing was achieved in the village of Manby, near Louth, twenty-five miles to the north.

Green's commercial instincts seem to have influenced his subsequent strange actions. No doubt he wished to return to Boston in an attempt, at least, to honour his promise to give tethered 'flights'. This latter state occurred when the balloon was not completely inflated as for full flight but still had sufficient buoyancy for lift. Long ropes, securing the craft to the ground and controlled by a steam winch, allowed the balloon to rise with its passengers to a height and for such a time that would simulate the sensation of flying. The process of ascent and descent could be repeated many times before buoyancy was lost and it became necessary to top up again with gas. Perhaps his thoughts were somewhat mercenary too, since by keeping the balloon semi-inflated he would save time and expense in re-filling it completely. In turn this would enable him to take up fare-paying passengers in quick succession and thus improve his earnings before nightfall.

Thus it was, that the strange spectacle of a flaccid balloon could be seen being conveyed in its semi-inflated state into Louth, 'for the purpose of gratifying Mr. Green's friends in that neighbourhood.' From there it was conveyed through the town as far as Burwell, ten miles along the main road to Boston. Here, it was reported, 'In consequence of the wind being excessively boisterous, it became quite unmanageable and was found necessary to deflate the balloon to avoid damage. Otherwise it was Mr. Green's intention to have conveyed it in a state of inflation to Boston.'

For Charles Green's next ascent the scene moves once again to Stamford. Thursday 7 September 1826 was the day chosen for his sixtieth ascent but due to extremely bad weather it was postponed until the following Monday.

During Green's visit to Stamford in 1825 he had promised a trip aloft for one of the more valiant among the local gentry. When the day of the event dawned it was declared that Mr Green would take up with him two passengers. One was to be Octavious Simpson, the son of Alderman Francis Simpson to whom Green had made his earlier promise. The other person was William Reed, son of the town-sergeant of Stamford. Blessed with favourable weather, their journey terminated in the village of Whittlesea near Peterborough. Octavious Simpson made copious notes during his trip of a lifetime, thus providing posterity with an excellent passenger's-eye view of a typical balloon flight of that era. This is his account:

'It was 4.08pm when the balloon lifted off from the gas works enclosure to the sound of cheers and clapping from assembled onlookers. As we rose, our course was easterly and I did not experience the giddiness I had expected. Instead I found the view of churches, streets and the profusion of buildings quite delightful.

Still rising, the balloon followed the silvery course of the river Welland. At a height of about 5,000 feet, air currents gently carried us in a south-westerly direction, over the twinkling spires and turrets of Burghley Mansion.

We passed over Barnack and left Pilsgate to the right. At 6,500 feet there was a splendid view of the panorama of the Fens, with an occasional glimpse of the sea between the clouds [this would most likely be The Wash]. On we went over Milton House, the Fitzwilliam's home near Peterborough, with Wansford and Castor away to the right.

After half an hour in the air the barometer had fallen from 29.7 inches to 23.3 inches, indicating a height of 8,000 feet. I found the panorama passing below me imposing beyond imagination.

Next a view of Thorney Abbey, with the ruins of Crowland Abbey in the distance, could be discerned, while the balloon itself approached the west end of that magnificent cathedral at Peterborough.

Mr Green began a controlled descent at this juncture so that we might gain a better view of the impressive architecture of the cathedral and to present the citizens of that city a good view of the balloon. The top of the cathedral seemed to be covered with human beings.

Having descended to 2,500 feet over the city, Mr Green threw out some ballast [on whom did it fall one wonders?] and the balloon once more rose steadily and Peterborough was slowly left behind.

The next object to capture my interest was Whittlesea Mere. The clouds were more dense at this point and bright sunshine danced playfully through them, reflecting upon the glassy surface of the lake. Now another straggling town, with much thatched roof architecture, hove into view. It contrasted sharply with the stone and tiles of the Stamford we left earlier. As the town drifted into view crowds of people could be discerned in the streets, waving and cheering. This was Whittlesea. The surrounding flat countryside seemed favourable for a landing, so Mr Green began a skilfully controlled release of gas, until the grappling hook on the landing rope began to find purchase and the balloon sank easily and gracefully to earth. It landed half a mile from Whittlesea at 5.15pm after a journey of eighteen miles.

Many willing people had ridden on horseback to the site and all helped to deflate and pack the balloon. Among the helpers were Mr Barber of Willow Hall, Whittlesea, Mr Maydell and Mr Wilson also from Whittlesea. Mr Green and I were then invited to partake of ample refreshment at the Haynes mansion in Whittlesea. A chaise and four conveyed us to Stamford which was reached at 9.30pm where we were met by a relieved crowd of thousands.

On this occasion, too, some indication is given of the cost of filling a balloon with coal gas. The *Stamford Mercury* declared it was:

> … happy to report the number of persons who paid to watch the inflation process in the enclosure (and who alone were the means to Mr Green of recompense for his outlay and professional skill) was larger than on former occasions. It is understood the Gas Company charged £10 for the exhaustion of the gas receiver [sic] and Mr Green's receipts were £88.

Inspired by Green's visit in 1825 and coinciding with his return the following year, the owners of a new public house built in 1826 in Blackfriar's Street and within sight of the gas works, were moved to name it the Balloon Inn. Sadly, by 1959, no longer used as an inn, the name had disappeared. In 1964, however, the property was renovated for domestic use and it is pleasing to record that, now converted to a delightful private residence, the present occupiers, having discovered the earlier name in the deeds, were moved to re-establish its connection with the past. Taking up residence in 1987, they renamed their home Balloon House.

Between September 1826 and May 1828 Charles Green made another thirty-two balloon ascents, one of which was at King's Lynn on 1 October 1827, before returning to Boston on Wednesday 14 May 1828. The ascent was made from Mr C.K. Tunnard's paddock in Bargate, with one passenger on board. It was, however, a most eventful flight, ending in the village of Gosberton, six miles north of Spalding.

Green's advertisement for the Boston event, his ninety-third, stated that no further trips were planned in that area for a considerable time. Although he returned to Stamford in August 1828 and continued to appear at other fenland venues occasionally, indeed he did not return to Boston until 1846, some eighteen years later. Charles Green's companion on this trip was Mr Willerton of Boston, and the pair lifted off from Bargate at 3 p.m., being

The Balloon House, formerly the Balloon Inn, near Gas Lane, commemorated Charles Green's first visit to Stamford in 1825.

buffeted by a blustery wind as they rose. After a while, though, Green evidently grew unhappy with the conditions. His experience and natural concern for the safety of his passenger in the now quite violent wind led him to decide to descend through the murk of the clouds to ascertain their whereabouts. This descent 'was with many perils'.

Nearing the ground in the vicinity of Gosberton village, still being bowled along by the boisterous wind, the grapple on the trail rope struck the tops of trees, breaking off some branches in its manic passage. It dragged a further half a mile, tearing at trees, hedges and buildings in its path like a demented snake. At one point it struck such a resounding blow on Mr Crosby's granary that the noise ' … frightened a labourer working inside and made him quite ill'. Eventually the grapple held in a cornfield but the effect of this violent decrease in speed was to eject poor Mr Willerton from his seat in the car. He saved himself from almost certain death only by clutching the carriage suspension ropes and dangling 150ft above the swaying ground, until the balloon finally came to rest. 'He displayed great presence of mind', remarked the local newspaper, in a nice piece of understatement!

With considerable difficulty, the balloon was deflated and the battered travellers were then free to sample local hospitality. Mr Everitt, mine host of the White Hart Hotel in Spalding, six miles away, but who happened to be passing the spot, generously gave up his seat in the post-chaise for Mr Green and his companion. They were conveyed to the nearby house of the Reverend J. Calthrop for ' … most gratifying hospitality', before returning to Boston by 9 p.m.

As has been mentioned earlier, the Green family produced a veritable clan of balloonists, and foremost of these after Charles was his brother Henry. Although Charles's son George also made no less than eighty-three ascents during his lifetime, there is only a single report of him visiting the Fenland region, and merely as a passenger to his father.

The one and only flying visit to the region by Henry Green, to Stamford in 1828, was clouded by a degree of public apathy very similar to that following the first flush of enthusiasm for powered flying nearly a century later, and in the words of the *Stamford Mercury*, 'The balloon ascent on Monday afternoon [18 August 1828] was not attended with any circumstances of particular interest.'

Only 450 people paid 2*s* (10p) each to watch the inflation process in the gas works enclosure so receipts were just £45. From this sum the owner of the balloon, Mr Brooker, had to pay the Stamford Gas Company for 19,000cu.ft of gas at the rate of 11*s* (55p) per 1,000cu.ft plus 'incidental expenses' for the use of the enclosure; altogether this amounted to £21: 'Various charges of other description attendant upon a speculation of this type … must have accounted to a sum which would leave precious little profit for Henry Green or Mr Brooker.'

Despite this inauspicious start Henry Green finally ascended from Stamford at 5.15 p.m. and with the aid of a brisk north-west wind he was carried rapidly in the direction of Burghley Park and its stately home. Henry himself now takes up the story:

> On account of the many clouds I was determined to remain as low as possible in order to gratify the inhabitants of this neighbourhood. Therefore I took up so much ballast that I had considerable difficulty in clearing some trees close to the place of ascent. These the balloon struck breaking off some branches, which dropped into the car itself. Quickly throwing out my flag and two bags of ballast, I cleared this obstruction and ascended rapidly. After a quarter hour I met a strong current to the north-west, which carried me from Thorney (my previous direction) towards Peterborough. However, before reaching that city I passed through a cloud and encountered yet another current which propelled me towards Whittlesea Wash [Mere] and thence to Wisbech.
>
> Another current verging towards the north carried me now towards Swaffham and I could see Downham Market off to the left. It was not long before I passed directly over the town of Swaffham having, at intervals, splendid views of Spalding, Wisbech, Cambridge, Ely and King's Lynn.
>
> At Swaffham I threw out more ballast and ascended considerably above the highest stratum of clouds. By now the sun appeared to be setting and knowing that a calm generally prevails at sunset, that circumstance, together with an apprehension that I was fast approaching the sea, induced me to begin my descent. I was, by then in a direct line for Ipswich, where I was anxious to land, to gratify my friends in that town and from which I had already made three ascents. In consequence, I discharged some gas and came below the lowest clouds from where I had a distinct view of Thetford, Brandon and Bury St Edmunds. I mistook the two former places for Ipswich and Woodbridge but in descending I noticed that the waters were so spread upon the lands by the late floods that I could not safely accomplish my purpose.

I then threw out sufficient ballast to check the descent and advanced at about the same height to Coney Weston, near Ixworth, twelve miles from Bury and eight miles from Thetford. My grapple struck the ground but dragged until the machine was checked finally when the car struck a tree. Owing to a deficiency of sand in Stamford, I chanced to take a four stone weight up with me for ballast. This I now found most useful, as by attaching it to the balloon, from rope it materially assisted the grapple.

Free ballooning is dependent upon the vagaries of the weather. Although these early aeronauts possibly took more risks in the course of their flights, as some of the above stories bear witness, it should be remembered that this was due more to their ignorance rather than sheer bravado. In mid-May 1830 Fenland skies reverberated to the thunder of horrendous electrical storms, accompanied by torrential rain and many were the frightening tales of 'bolts from the sky'. Heavy rain persisted until Wednesday 26 May, the day of Charles Green's 149th aerial voyage, when in Peterborough the morning dawned fine. From the number of gigs and carriages bringing people into the city, it seemed it would soon be full to overflowing. Spirits were somewhat dampened in the afternoon, however, by a return of squally rain showers delaying the balloon launch until 7.15 p.m.

As ever, Green was keen not to disappoint his audience despite the inclement weather and at that hour took his seat in the car. As a precaution against the conditions he had filled the balloon to only two thirds its capacity and consequently was obliged to leave behind a gentleman who was to have accompanied him. To strains of 'God save the King' from the town band the balloon rose gently, watched by a much-depleted crowd of 200 in the gas works enclosure and by about 1,000 others in the surrounding roads.

Taking up an easterly course the balloon flew towards Wisbech, but with the wind constantly changing direction Mr Green deemed it wise to descend. He landed in Littleport Fen, about three miles from March. Noting this event in his diary, farmer John Peck wrote:

26 May 1830. Went to [Wisbech] Horse Fair; a poor concern; home by dinner. Round Parson Drove in afternoon to see men topping wheat. About 7 [p.m.] in evening Mr Green's Balloon passed over to the south of Parson Drove, from Peterborough. Very plain to be seen.

In view of the most atrocious weather during the previous few weeks, Charles Green decided to take advantage of its turn for the better and give the people of Peterborough another opportunity to see his balloon. He therefore announced a repeat performance for Saturday 6 June 1830. There was some reluctance for rejoicing and merrymaking at this time as it was being widely reported in the press that King George IV's death was imminent. From 6 a.m., though, the city was a continuous scene of bustle and noise from innumerable vehicles of all descriptions, flocking into the town from all directions. On this, Charles Green's 150th voyage, his brother George and Mr H. Miller, a resident of Peterborough, would accompany him.

With brilliant sunlight setting off the magnificent richness of the colours of the balloon, watched by 1,000 paying spectators in the gas enclosure, the aeronaut and his two companions rose into a cloudless sky at 6.40 p.m. As they gained height the travellers estimated the city streets around the enclosure to be teeming with up to 20,000 onlookers gazing up at them. Loud and continuous cheering echoed skywards and could be heard for many minutes. For a detailed account of this journey we can turn to Mr Miller's own recollections:

After leaving the ground the balloon took a north-east direction, passing the village of Eye at a height of 4,000 feet. As we rose further a new current was found, at about 9,000 feet, which swung us more to the east but not sufficiently to reach Wisbech. Mr Green (senior) thus decided to descend into the earlier air current and seek a course in the general direction of Spalding. In order to find this current we descended to 2,000 feet, two miles east of Crowland and close enough to the ground to be able to converse distinctly with the local peasantry, who tried to tempt us to land at that place. Shouting that we were bound for Spalding, Mr Green threw out some ballast and we soon outran them.

Drifting along in the gathering dusk, now having reached an elevation of 17,000 feet, it seemed to me that turnpike roads below were no wider than the mere track of a carriage wheel; the sheep in fields appeared as maggots crawling over a cloth; cattle as small dogs and vehicles on the roads like children's toys, only identifiable by their motion. At our highest elevation I experienced a temporary deafness to such an extent that I was hardly able to hear a word uttered by Mr Green until we neared the ground once more. This did not, however, prevent me joining my companions in drinking hearty toasts for 'His Majesty's restoration to health' and 'All friends below.' It was our real intention to reach Spalding but the lateness of the evening, in the end, prevented our doing so. Mr Green was doubtful that there would even be sufficient time to pack up the balloon without mishap before darkness fell.

After a most pleasant journey of two hours and perceiving the sun to have set, we determined to land near the approaching village of Moulton Chapel, some four miles from Spalding. To our delight we were greeted and assisted not only by the local inhabitants but also by some gentlemen who had ridden after the balloon from Peterborough. An invitation was received from Richard Morton Esquire of Peak Hill, [Cowbit] to proceed to his mansion where we were met with the greatest of hospitality. Having slept there and partaken of a sumptuous breakfast the following morning, a chaise and four was obtained from Spalding enabling us to return to Peterborough at 11am on Sunday.

That erstwhile observer on the ground, John Peck also watched the balloon's progress:

Saw Mr Green's Balloon, it ascended from Peterborough and took a direction straight for Parson Drove. When over Thorney the wind veered to the south and took it near Moulton Chapel where it alighted after being up upwards of two hours, during the whole of which time it was visible at Parson Drove.

Well satisfied with the commercial and public success of his latest venture, Charles Green left Peterborough the following Tuesday bound for Lincoln, the venue for his next ascent. For this, the first manned balloon ascent in the city, the *Lincoln Herald* carried an advertisement in its 2 July issue:

ROYAL BALLOON.

UNDER THE PATRONAGE OF THE DIFFERENT MAGISTRATES OF THE COUNTY,

AND THE

MAYOR & CORPORATION *of the* CITY *of* LINCOLN·

POSTPONEMENT

OF THE ASCENT

OF MR. C. GREEN'S BALLOON

AT LINCOLN.

At the Request of the Mayor and Corporation of this City, and from consideration of public propriety, Mr. C. Green has been induced to POSTPONE his Ascent from Lincoln, (*on account of the lamented death of the King,*) until MONDAY NEXT, JULY 5th, at Four o'Clock in the Afternoon.

MR. CHAS. GREEN the Veteran Aeronaut, Patronised by His Most Gracious Majesty, at whose Coronation he made his first ascent, respectfully notifies to the inhabitants of Lincoln and its vicinity, that he purposes making his 151st aerial voyage, on Monday next July 5th, 1830, at 4 o'clock, from a spacious Inclosure near the Gas Works, belonging to J. B. Cottill, Esq., and in the occupation of John Allison, Esq., both of which gentlemen have liberally granted the use of the said Inclosure on the occasion. Tickets of admission to witness the Inflation and Launching of the Balloon, 2s. each, may be had at the Library, the principal Inns, and at the Booksellers.

Advertisement in the *Lincoln Herald* for Charles Green's balloon exhibition in the city on 5 July 1830. (Lincoln Library)

At the request of the Mayor and Corporation of this City and from the consideration of public propriety, Mr C Green has been induced to POSTPONE his Ascent from Lincoln (*on account of the lamented death of the King*) until
MONDAY NEXT, JULY 5th

Midsummer fair day in Lincoln dawned fine with blue skies and little breeze and on this occasion his son George, who was making his own forty-second flight, accompanied Charles. Inflation of the *Royal Coronation* balloon required 142,000 gallons (23,600cu.ft) of coal gas, a process overseen by Mr Needham, the gas works manager. It began at 11 a.m. in a paddock in Newland and could have been completed in as little as three hours but took longer due to Green's astute commercial desire to let as many paying customers as possible see the actual process for 2s (10p) each. By the time of the launch at 5 p.m., 800 visitors were in the enclosure being entertained by the Lincoln Band. It was noted that the inflation process was 'by the improved plan of elevating the head of the balloon by ropes attached to a pole of considerable height, before the filling commenced, the old London bungling [sic] mode of swelling it out on an extended platform, is superseded',

After a short speech from the crimson silk-covered car, the clock struck 5 p.m. and Charles cast off the last rope amidst loud cheers from all sides. These were echoed as the balloon rose gently over the heads of thousands of people watching from vessels on the river, its banks and along the whole brow of the hill from the asylum to the hospital. At 500ft Green released a small parachute with a basket containing a cat, which landed safely in the paddock below. At first the balloon flew north, almost over the west end of the cathedral until, following the Brigg road and, still climbing, it encountered a current that swung it eastwards.

Now the aeronauts could see the whole panorama of the Lincolnshire coast before them, from the Humber to Skegness and on into The Wash, with the Norfolk coast visible in the distance. Over the bend in the river Witham, at Grubhill, a fresh current took the balloon south-east and Charles was now hopeful of reaching Boston. Nearing Tattershall, at an altitude of 12,000ft, yet another change of wind direction set them north of east towards Horncastle and so, having been aloft for an hour and forty minutes, and with the prospect of reaching the sea, Green began to vent gas for a descent.

It was 7 p.m. as many willing hands converged on the balloon, catching the grapple and guide ropes to help bring it to earth in a pasture at Scrivelsby Park, the seat of the Hon. the Champion Dymoke, where the gas was discharged entirely. In the absence of Mr Dymoke, his steward arranged a horse and gig to carry the two aeronauts and their balloon into Horncastle, some two miles distant. Such was the warmth of the welcome and hospitality they received in Horncastle that their departure for Lincoln was delayed until midnight. It was 3 a.m. the next morning before they reached the city where, with but a few hours respite, they were again fêted at breakfast – heroes of the hour!

Prior to this particular flight, Charles Green had encountered an obstructive response to his original application to make the ascent from Lincoln Castle Yard. Despite his sponsors being all the county magistrates – who themselves recommended approval – the sheriff (custodian of the Castle Yard grounds) refused permission, and subsequently the event was moved to Newland.

This spat with authority did not prevent Charles from staying on in Lincoln to make another ascent, his 152nd, from the same venue, on Tuesday 13 July 1830. This trip was postponed from Friday 9th due to high winds and a lady, 'well known in the neighbourhood', who was billed to accompany Charles, changed her mind on the day so Mr J.B. Cuttil, in whose paddock the event was held, took up the offer of a flight.

It was to be another long day for the intrepid aeronauts. They took off at 6.30 p.m. and were carried north-east to land safely at Rothwell, four miles east of Caistor after a journey of one hour and forty minutes. Sumptuous hospitality once again held them in Caistor and it was 6 a.m. next morning when they arrived back in Lincoln. An aeronaut certainly needed plenty of stamina, a fine appetite and a robust constitution in those days!

James Green, another of Charles's aeronautical brothers, no doubt needed a robust constitution when he was thrown from his balloon car during a landing on 18 June 1830. His demise came to light when a balloon with a 10ft tear in the lower part of the fabric and an unmanned car, passed over Newborough before coming to earth in Thorney Fen, near Peterborough. The grapple iron was missing and its rope had parted close to the hoop, indicating the ferocity of its earlier landing. The balloon and equipment was gathered up and stored in a certain Mr Perkins' barn until extensive enquiries established that it belonged to James Green, who had taken off from Coventry.

Two years were to elapse before Charles Green, or for that matter any other aeronaut, graced the region with his charismatic presence and it was to King's Lynn that he next came to celebrate and advertise the opening of a new gas works, near the town's South Gate.

Believed to be associated with that event, a small, unmanned, advertising and 'wind-test' balloon caused quite a stir in Gosberton Fen, near Spalding on the evening of 20 August 1832. According to a report in *The Bee* (otherwise known as *Stamford Herald and County Chronicle*, a short-lived rival to the *Mercury*):

> An elegant silk balloon, about the size of a hogshead barrel, bearing the name GREEN, descended on the farm of Mr Jacob Smith. It was first discovered by a shepherd who, seeing so rare a sight amongst his master's flock, set his dog upon it. The shepherd then struck it with his spittal, whereupon two holes were made and the simpleton was near suffocated by the gas it contained.

This incident bears remarkable similarity to what must have been one of the first – if not the very first – sighting of a balloon in the East Midlands almost fifty years earlier. On that occasion, however, it was reported in a much more charitable tone than this later one. The *Cambridge Chronicle* of Saturday 19 February 1785 carried the following story:

> On Monday the 9th instant, a balloon was launched at Soham in this county, which ascended very majestically until it lost itself in the clouds; and falling in the fields between Fordham and Newmarket, was observed by a man and a boy. The lad forsook his work and fled precipitately from so tremendous a sight, imagining it was something sent from heaven to visit him for his past transgressions; the man, whose curiosity so far prevailed over his fear as to enable him stand his ground, stood aghast, and in a summary way took retrospective view of his past life, and in finding his conscience clear (the only true source of courage) ventured to approach it and said, 'In the name of the Father, Son and Holy Ghost, what are you? Speak and tell me.' But, at length, finding it speechless and motionless, he ventured to take it up and carry it home.

In his aeronautical career, which had by 1835 spanned an exciting fourteen years, Green was averaging more than one trip a month. He had taken all manner of risks, some of which were quite calculated, in the furtherance of this branch of science and had emerged unscathed. This is a credit to his single-minded but disciplined approach to his work. By the mid-1830s Green's reputation was unrivalled and he seems to have reached a stage in his career where he needed fresh challenges. His attention was drawn to long distance flights that required an understanding of the meteorology of flying and in particular the patterns and use of air currents. It was, perhaps, not surprising therefore to read in the newspapers of September 1835 the headline 'Mr Green the Aeronaut Lost.' Had fate caught up with Charles Green at last? This was indeed the question on everyone's lips and as one national newspaper put it:

> Considerable anxiety was on Friday [18 September] evinced by the public for the fate of this gentleman. Frequent enquiries were made of the proprietors of The Vauxhall Gardens

[London] as to where Mr Green had descended, to which no reply could be given, they [the proprietors] not having heard of him since his re-ascent at Walthamstow.

Being, by now, something of a national celebrity, Charles Green commanded this degree of anxiety but on this occasion Green himself had a considerable hand in creating an aura of mystery about this, his latest aerial expedition in his trusty *Royal Coronation* balloon. After some days the final story emerged and its outcome was clearly wrapped up with his growing desire to undertake more adventurous balloon journeys. This one was to bring him once again to the fenland region.

That Thursday evening (17 September) saw Charles at the Vauxhall Gardens, situated on the South Bank of the Thames near Blackfriars Bridge – a regular venue for his balloon exhibitions. The balloon was already inflated and a germ of an idea was no doubt already in his mind, for he was not a fellow to go off 'at half cock'. Chatting with a committee member of the Floricultural Society, he remarked enigmatically that he sought 'a gentleman of light weight but with a heavy purse' to accompany him on this flight.

Bearing in mind it was already mid-September and approaching 5.30 p.m., there would be little daylight remaining to complete much of a journey that day. However, his remark was taken seriously and immediately communicated to one Mr Butler, a surgeon from Woolwich, known both as a gentleman of means and keen to undertake a flight, a combination seemingly less prevalent than might be imagined. Mr Butler must have been close by, for having been introduced to the aeronaut – to whom he offered the princely sum of 10 guineas (£10.50) for the privilege of a flight – the two were airborne by 6 p.m.

To the accompaniment of loud cheering along its course, Green's balloon was carried north-east on a gentle breeze. Crossing the Thames above Blackfriars Bridge it gained height steadily and within half an hour had reached the parish of Walthamstow, with the city receding to the south and the countryside opening out below. It was at this point in the journey that events took another intriguing turn.

Mr Green brought the balloon to earth in a field on the north-east side of the town where, according to Mr Butler's version of events, he was asked to alight from the car and make his own way back to London. Mr Butler claimed he expressed his extreme disappointment to Mr Green at not being permitted to remain for the whole voyage and was told, 'It is quite impossible.' Upon enquiring why, Mr Green replied enigmatically, 'I cannot tell you now but will on another occasion and you will be satisfied.' Pondering on that assurance, the bemused Butler watched the aeronaut take on board a small quantity of ballast and promptly ascend again into the gathering dusk, still heading north-east.

Mr Butler returned to Vauxhall Gardens to recount his strange tale and naturally there was much speculation as to Mr Green's fate. It emerged that prior to his last balloon ascent he had informed the proprietors of the Vauxhall Gardens, Messrs Gye and Hughes, that if the wind was favourable he would attempt to cross the Channel and descend in France. His plan was thwarted at the last moment by a change of wind direction and he had to abandon the idea on that occasion.

Listening to Mr Butler's story, it was, in the opinion of these gentlemen, quite possible that Green had decided to make another attempt at a sea crossing but wished to play down the matter in case of another failure. They knew that in the car there was only a pint of wine and a few biscuits, and surmised that such meagre provisions for a journey of any distance and having so keenly taken a companion – with whom he then abruptly parted company – indicated the possibility of undertaking a long journey was a sudden decision. Butler, adding that Mr Green had informed him they would land at Walthamstow 'for a particular purpose', reinforced this conjecture. The conclusion reached by these anxious gentlemen was that if the all-important wind did not change, the course of Green's balloon must have been across Suffolk, Norfolk and the German Ocean [North Sea] to make a landing in Holland.

In the event, Charles Green was quite safe and sound and in no danger whatsoever, but there was great relief when he finally turned up at Vauxhall Gardens where he was able to give his version of events:

> On reaching Walthamstow I informed Mr Butler of my intention to remain aloft all night. He was most anxious to remain with me to savour the experience but I did not deem this prudent. It was my opinion that in the event of being carried out to sea during the night it would be much more advantageous to have his weight in disposable ballast. Having taken on board a fresh supply of sand I re-ascended and continuing northwards, passed over Bishop's Stortford, Royston and Huntingdon. I encountered an air current which took me more to the east and it was by now quite dark. Below me, those towns lit by gas illuminated the atmosphere for a considerable distance and were easily distinguishable. I fired several Bengal lights during the night and from shouts reaching me from the darkness below, doubted not that their appearance must have created much astonishment.
>
> It was 6am on Friday morning [18 September] when I decided to descend in the parish of Wimbotsham in Suffolk. Though it was an early hour I was soon surrounded by a large crowd, all anxious to render me assistance. In response to his kind invitation, I partook of an excellent breakfast at the residence of Mr J Pike, after which as the sun's rays having evaporated overnight dew from the balloon and the gas once more expanded, I rose into the air yet again.
>
> I hoped to find either an inland current or one that would take me across the Channel. However, perceiving that I was going rapidly towards the German Ocean [North Sea], I allowed the gas to escape. The balloon came safely down to anchor within one mile of the sea and close to the town of King's Lynn, at 10.30am. While deflating the balloon and packing it up I received numerous congratulations, though upon hearing my story many people, I believe, were astounded by it and doubted my claim to have been in the air this past night.

In fact, Charles Green upon his safe return to the Vauxhall Gardens in London cleared up the mystery that had surrounded his 'disappearance', stating he remained in the air for fifteen hours overnight. What Green did not tell his friends, though, was that making a long distance flight with an overnight component was his practice run for the greatest aviation journey of the century.

Charles Green, now fifty years of age, turned his efforts to a number of scientific and exploratory balloon journeys during the next decade, and for this work he used a new craft named *Royal Vauxhall*, a 70,000cu.ft monster, 150ft in circumference and 80ft high, made from finest silk at a cost of £2,100 to its owners, Messrs Gye & Hughes,

THE VAUXHALL ROYAL BALLOON.

Charles Green's 'Royal' balloon carrying nine passengers aloft from the Vauxhall Gardens, London, on 9 September 1836.

proprietors of the Vauxhall Gardens in London. Green's new programme included long-distance flights in Britain and on the continent; altitude records; flights in adverse weather conditions; involvement with the ill-fated parachute experiments of Robert Cocking, and not least, the Great Flight to Germany.

The latter flight from London to Weilberg near Koblenz in Germany made on 7–8 November 1836 rates as one of the classic balloon journeys of all time. Robert Holland, MP for Hastings, with publisher and theatre impresario Monck Mason, provided the project funding and both men accompanied Green on the flight. Much has been written elsewhere about this famous air journey so suffice it to say that taking off from the Vauxhall Gardens at 1.30 p.m. on 7 November, the intrepid aeronauts crossed the English Channel, France and Germany to land in the Duchy of Nassau near Koblenz, eighteen hours later, having travelled a distance of 480 miles (772km). In honour of the event Green re-named his balloon *Nassau*, and the distance flown stood as a record in Europe until 1907.

It was, therefore, more than ten years before the aging but still revered Green would return to the fenland region, when Great Britain had entered the Victorian era with the new Queen's accession in 1837. The following year the coronation of Queen Victoria was celebrated in great style throughout the land and not least in the fenland town of Wisbech. The town square was a rare sight to behold, with row upon row of tables and benches for revellers to eat their fill. On Thursday 2 July 1838 more than 5,000 people sat down to eat, with many more watching from every window and balcony overlooking the scene. To cap this scene, a brilliant blue and white striped balloon ascended from near the market place and descended a mile from the town. Unfortunately the name of the aeronaut involved is not recorded, but the spectacle was repeated the next day.

If one wished to mark a great occasion in fine style, then there was still nothing to beat a balloon for sheer spectacle. If, too, one wished to have the best aeronaut for such a celebration, then Charles Green was the obvious choice. Now in his sixtieth year, Green, veteran of more than 300 ascents but with hundreds more still to come, returned for his last trip from Stamford on Thursday 30 April 1846.

There was to be a great festival at Burghley House in celebration of the coming of age of Lord Burghley, eldest son of the Marquis of Exeter, and Green was to be a star attraction. A grand dinner and ball was planned, '… but for the amusement of those who did not wish to participate in the dinner or ball, Mr Charles Green was engaged to ascend in his balloon, the *Albion*, from the gas works in the town'. Extending his generosity still further, the Marquis distributed among his cottage tenants and the poor people of Stamford and neighbouring parishes a total of 6,500lb each of beef and bread, 3,500 gallons of strong ale, and a gratuity in money according to the size of each family. It was indeed a most memorable occasion.

Accompanied by Mr Jones, the Stamford gas engineer, Charles Green lifted off on his 319th aerial journey at 6.30 p.m. By good fortune the balloon's course took it to the south-east and almost over Burghley House itself, but the remainder of the flight appears to have been quite uneventful. Having passed over Peterborough, Mr Green descended safely at the village of Oxney, three miles to the west of the city, at 7 p.m.

For his final excursion into East Midland skies, Charles Green made the sort of journey for which he will always be remembered, described in his own words as 'a quiet airing and very pleasant ride with a companion across country'.

It was at 7 p.m. on Tuesday 4 September 1849 that a balloon was seen rapidly descending from a considerable altitude in the vicinity of West Newton, near King's Lynn. Curious local inhabitants were drawn like moths to follow the monster as it was borne along by the wind. With the assistance of several strong harvestmen the balloon finally came to earth in a meadow farmed by a Mr Griggs, in the picturesque village of Appleton. Enquiries established that the craft was the property of 'Mr Green, the veteran and well-known aeronaut accompanied by an elderly gentleman', who was in fact his long-time friend George Rush of Elsingham Hall, Essex; they had left the city of Norwich two hours earlier.

A veritable army of willing helpers from the village, under the eagle eye of Mr Green, deflated, packed and stowed the balloon on a farm cart and the pair made their way to King's Lynn and thence back to Norwich. *The Lynn Advertiser & West Norfolk Herald* took great care to report that 'the utmost decorum was observed by all parties and little or no damage done to fences or to the balloon by those whom curiosity had drawn to the spot'. The *Norwich Mercury* published a more informative report that, although containing

Charles Green and his *Nassau* balloon carrying six passengers over Norwich on 24 September 1840.

Advertisement for Charles Green's display of his *Nassau* balloon in Norwich in September 1840. (Ray Wilson)

conflicting dates, is all the more interesting for the insight it gives into the purpose of this flight. It particularly reinforces the important role played by Green in organising and piloting balloon ascents so that scientists and astronomers – both amateur and professional – could use the valuable opportunity that balloon flight gave to the study of the upper atmosphere. Where Charles Green went, Henry Coxwell could only but follow.

George Rush was a wealthy amateur scientist and astronomer who first met up with Charles Green back in 1838 when the former wanted to test his experimental aneroid barometer, and the latter was considered the best man to organise some high altitude balloon ascents by which to conduct the tests. Thereafter, Rush made thirteen trips with Green, albeit as his sponsor (Green charged Rush £300 for that first flight!), but equally they collaborated as friends, whose aeronautical prowess on the one hand and scientific interests on the other ideally complemented each other.

Nine years had elapsed since Charles Green last brought Norwich to a standstill with the awe-inspiring sight of himself and no less than seven local dignitaries ascending majestically into the sky beneath the great *Nassau* balloon on its way to Metton. The primary purpose of this latest journey in 1849 was 'to test the powers of a newly-invented instrument called the aneroid barometer with the common [fluid] barometer'. The *Norwich Mercury* continued:

> The balloon used was not the one used for general ascents but a smaller one called *Victoria* with a capacity of 20,000 cubic feet only. Filling began during the morning at the Barrack Yard [in Norwich] and the time of departure was to be 5pm. Although it was not in the nature of a general exhibition, a payment of one shilling [5p] was requested of persons who visited the Barrack Yard. The comparatively few who availed themselves of the opportunity to rub shoulders with the huge rolling, spinning sphere must have convinced Mr Green that, without some attractive adjunct, the ascent of a balloon is not itself sufficient to collect a large company together in Norwich.

This latter situation was due to the canny folk of Norwich flocking to all the vantage points around the city to gaze upon the scene free of charge! St James' Hill nearby was, for example, covered by a dense mass of spectators from its base to its apex, and it was likewise along the crest of Mousehold. Nevertheless, children from the Hospital School and the Union House were admitted to the Barrack Yard to watch the balloon being inflated:

> Gas was supplied to the balloon by a small main brought through the wall of the barracks into the barrack ground. When inflation was complete, Mr Rush took his seat to tuneful music from the band of the 16th Lancers, Mr Green operated his quick-release lever on the restraining rope and the huge sphere leaped into the air. Rising almost vertically the balloon headed west by south and soon melted from view in heavy clouds, only to reappear about fifteen minutes later still on the same course.
>
> The descent took place at West Newton in a field about four miles from the sea [The Wash] near the home of Sir W B Folkes, Bart at 7pm. Both aeronauts then travelled to [King's] Lynn from whence Mr Green travelled by railway to London on Wednesday morning and Mr Rush returned to Norwich.

There then followed a table of Mr Rush's instrument readings and observations for the voyage during which they reached an altitude of 12,226ft in relation to their take-off point:

ASCENT

Barometer	Thermometer	Aneroid	
30.20in	74° F	29.74in	'moment of quitting the earth.'
29	68	29.70	
28	66	28	
27	65	27.50	
26	64	26.50	Here Mr Rush shook the aneroid greatly, the effect of which was that it fell more rapidly.
25	63	23	
24	61	24.50	
23	61	24	Here the aneroid ceased to act.
22	54		
21	52		At this period the time was 5h. 45m.
20	52		Time 6h. 10m.
19	46		

At the greatest elevation one of the bottles was emptied of the water it contained and the stopper carefully replaced for the purpose of securing air from the upper region.

DESCENT

Barometer	Thermometer	Aneroid	
21	46		Here the balloon revolved considerably upon its axis in consequence of the rapidity of its descent.
24	48		Here the aneroid began to act again and the second bottle was filled with air.
26	50	25.50	
27	51	26.50	
28			Here Mr Rush put up his instruments and discontinued his observations.

The reason for the aneroid barometer 'cease[ing] to act' is that it most likely had a limited range that would make it unsuitable for measuring at heights above 4,000–5,000ft but would start to 'act again' when their descent came below that height.

Having, in his remarkable career, conveyed a total of over 700 passengers aloft during the course of 526 flights, Charles Green, the nineteenth century's greatest aeronaut, made his last flight on 13 September 1852. His final few flights were in the company of John Welsh of Kew Observatory and confirm that Green, contrary to some historians' opinion, embraced more than once the role that a balloon could play in facilitating high altitude scientific observation.

Charles retired to the most aptly named Aerial Villa in Holloway, London, but sadly neither his birthplace nor his retirement home exist any more, and the only tangible

THE LATE MR. CHARLES GREEN, THE AERONAUT.

The 'old ethereal flier', an engraving of Charles Green made shortly before his death in 1870. (Spalding Gentlemen's Society)

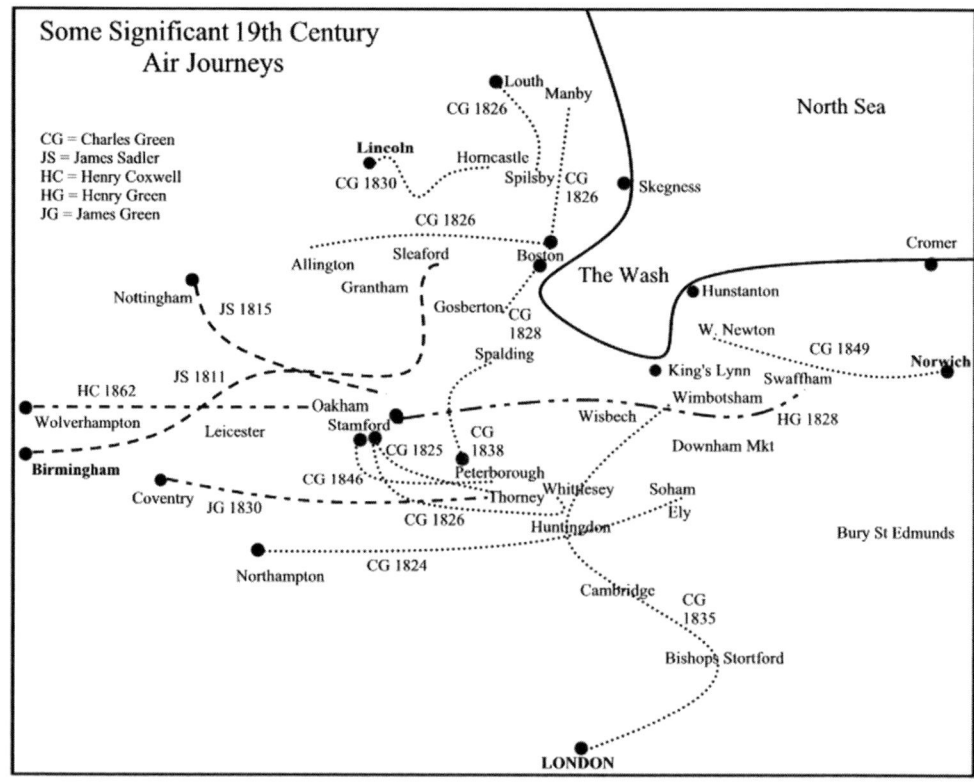

Some Significant 19th Century Air Journeys

CG = Charles Green
JS = James Sadler
HC = Henry Coxwell
HG = Henry Green
JG = James Green

Louth
Manby
CG 1826
North Sea
Lincoln
Horncastle
CG 1830
Spilsby CG 1826
Skegness
CG 1826
Cromer
Allington
Sleaford
Boston
The Wash
Grantham
Hunstanton
Nottingham
JS 1815
Gosberton
CG 1828
W. Newton
CG 1849
Norwich
Spalding
King's Lynn Swaffham
HC 1862
JS 1811
Oakham
Wimbotsham
Wolverhampton
Leicester
Stamford
Wisbech
HG 1828
CG 1825
CG 1838
Downham Mkt
Birmingham
CG 1846
Peterborough
Coventry
JG 1830
Thorney Whittlesey
Soham
CG 1826
Huntingdon
Ely
Bury St Edmunds
Northampton
CG 1824
Cambridge
CG 1835
Bishops Stortford
LONDON

Map showing some significant balloon journeys in eastern England during the nineteenth century.

monument to him is an ornate grave marker in Highgate cemetery, where he has a richly deserved plot in that final dwelling place of the great and the good. This 'Grand Old Man of the Air' passed away quietly on 26 March 1870, aged eighty-five years, having been born just a few months after Lunardi's first balloon flight in England.

Between the years 1840 and 1850, the approach to the science of ballooning, as portrayed by Sadler and Green for example, underwent a gradual but significant change. With a few notable exceptions – such as Gypson and Coxwell – gone were the relatively long flights, each of which was regarded as an individual voyage of discovery and epitomised by these pioneers. Ballooning was now entering the Victorian age of showmanship.

CHAPTER TWO

Showmen to Sportsmen

Richard Gypson's arrival on the Victorian ballooning scene in 1839 was an indication that the world of aeronauts and aeronautics was beginning to take on a more devil-may-care form. From that time, too, there can be distinguished a significant change of emphasis in ballooning events that had become, and would continue to be, a feature of Victorian leisure time. Principal among those changes was the proliferation of aeronauts themselves; the appearance of a few charlatans; the notably shorter distances flown and exploitation of the potential dangers of parachuting for financial gain.

Having twice entertained spectators in Northampton during 1840, Richard Gypson moved on the following year to beguile audiences across the Fens with no less than six separate balloon exhibitions during 1841. First among these, his own thirty-ninth ascent, was in Wisbech on Monday 26 July 1841. 1,700 people paid 1s (5p) to enter the enclosure in which his *Royal Standard* balloon, of 118ft in circumference and 20,000cu. ft capacity, would be inflated. This process took place in a field opposite the gas works on Leverington Road where, at 6.30 p.m. on a fine summer's evening, the balloon, emblazoned with alternate crimson and gold stripes, was released.

For this ascent Gypson was accompanied by Mr G. Robins, a native of Wisbech and a kitten; the latter, no doubt, being somewhat frightened by its demise! Amid shouts from an immense crowd in the surrounding area 'the large balloon rose in majestic style.'

Reporting the event, the *Stamford Mercury* described the remainder of the flight:

> When at a great height Mr Gypson let fall a parachute containing a kitten. It [the kitten] descended slowly and steadily [and safely]. The Balloon was out of sight for a short time and was found to have made an excellent landing at Cotton's Common on the Downham Road, near Upwell, after a journey of six miles.
>
> The Mayor of Wisbech kindly lent his carriage for the pursuit of the Balloon and it was on the spot very soon after the landing. Messrs Gypson and Robins deflated and packed up the silken balloon and were driven in the carriage through cheering crowds as they rode in style through the streets of Wisbech.

'At the earnest request of several Gentlemen, Mr Gypson is induced to make a second ascent with his balloon', and seizing his opportunity, an event was arranged for Thursday 12 August 1841, the day of the local Bullock Fair. At 7 p.m. on that balmy summer evening, Mr Gypson lifted off from the same field. His companions this time were Mr R. King, the son of Reverend King of Wisbech; Mr M. Marshall, son of the gas works manager; a small puppy and a kitten, the latter animals belonging to a Mr Thirkell, also of Wisbech. Secured inside a container, possibly a wicker basket attached to the parachute, these hapless animals were duly dropped safely as the balloon rose into the clear blue sky. After an uneventful flight of fourteen miles the three gentlemen descended at Bexwell, near Downham Market.

Richard Gypson had an eye for business and soon increased his earnings by writing accounts of his aerial journeys for local newspapers. He was blessed with a very fair turn of phrase and his accounts are quite entertaining if somewhat flowery. The *Norfolk Chronicle* and *Norwich Gazette* of 21 August 1841 provide an example of his literary prowess:

> Thursday morning was ushered in upon us with all the ominous aspects of being like many of the dreary, cheerless, wet and windy days that had preceded it. As the morning advanced and the sun rose, the wind evidently increased until the sheets and cloths forming the enclosure were in many places hurled from their fastenings. Several of my friends seriously persuaded me to postpone the ascent but having assured some country gentlemen who formed a committee for their district, that if the violence of the wind did not actually prevent the ascent by causing something disastrous to the Balloon, it should not prevent it as far as I was concerned. I felt bound to keep my word and commence the inflation, which began before one o'clock

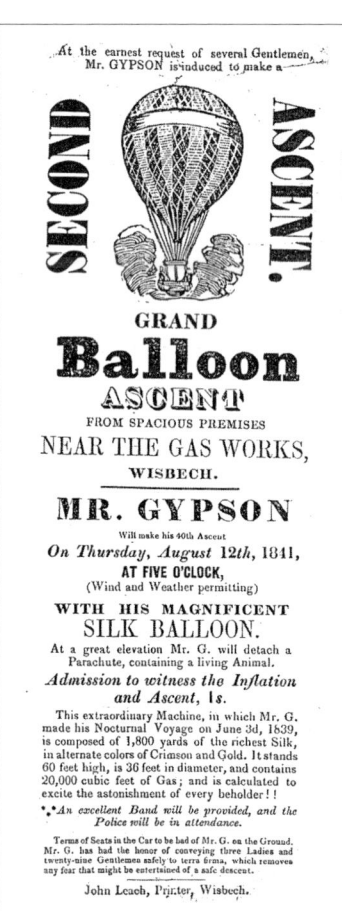

Advertisement for Richard Gypson's balloon exhibition at Wisbech gas works on 12 August 1841. (Ray Wilson)

amidst the howling of the wind, the falling of poles and the drifting of cloths and canvas that surrounded us. The balloon itself, as the gas passed through it, rolled and plunged to and fro like one of the monsters of the deep on the surface of an ocean. I was compelled repeatedly to stop the passage of the gas, in order to arrange the silk properly within the network. Most providentially, as we got deeper into the afternoon, the wind subsided and the latter part of the day was altogether as calm and as fine as the morning had been boisterous and unpleasant.

All the arrangements being completed and my companions [Messrs King and Marshall] having entered the car, I gave the signal for the ascent. My balloon rose in fine perpendicular order and instantly a flood of light and beautiful scenery burst forth with the country brought immediately into view, all apparently receding. The scene was lit up into magnificence by the refulgence [sic] of an evening sun, while we were suspended in the atmosphere verifying the expression of the poet:

Between two worlds hoves like a star
Twixt night and morn upon the horizon verge.

My two companions were enrapt at the splendour of the prospects around and beneath us, the evening being fine and clear the sight of the Ocean [possibly The Wash] was beautiful beyond everything. I never, in any ascent, saw anything so magnificent and superb as the splendid prospects it offered to our eyes. Tinted by the crimson rays of the setting sun it looked like an immense mass of bright vermilion, here and there studded with some dark objects as black as jet, doubtless the vessels floating on its surface. Whittlesea Mere was also distantly visible and had an equally delightful appearance. Now we had reached a very considerable elevation and the expansion of gas was rapid and powerful, causing me to part with great quantities by means of the upper valve and somewhat checked the upward course.

As we occasionally paused in our conversation, the universal quiet that reigned all around was very solemn. Nature was entirely noiseless, even the wind was silent.

With night now advancing upon us, the loss of the sun's rays caused a condensation of the gas and we found ourselves descending. The Ocean had changed its crimson tint for one tinted with the darker shades of night as we made for Mother Earth. Crossing the [river] Ouse with a very steady descending power till the grappling iron caught hold in a piece of grazing land in the parish of Bexwell. We came finally to ground in a turnip field in the parish of Wimbotsham, where Mr Muskett, the owner, was very hospitable. He forwarded the Balloon to Downham Market, whither we proceeded later. Here we procured a chaise and set off for Wisbech to be greeted by Gentlemen who had pursued us [on horseback].

My companions behaved coolly and were much delighted by their aerial trip. I am also most grateful to the Gas Manager and the Police under Superintendent Rust.
Signed: R Gypson.

John Peck, the gentleman farmer from Parson Drove mentioned in the previous chapter, was also in Wisbech that day and he stayed to watch the proceedings, noting in his diary, 'In the evening saw a Balloon go up carrying the proprietor [Gypson], Mr King and Mr Marshall. The air being clear, it was in sight for half an hour and descended about a mile beyond Downham, Norfolk.'

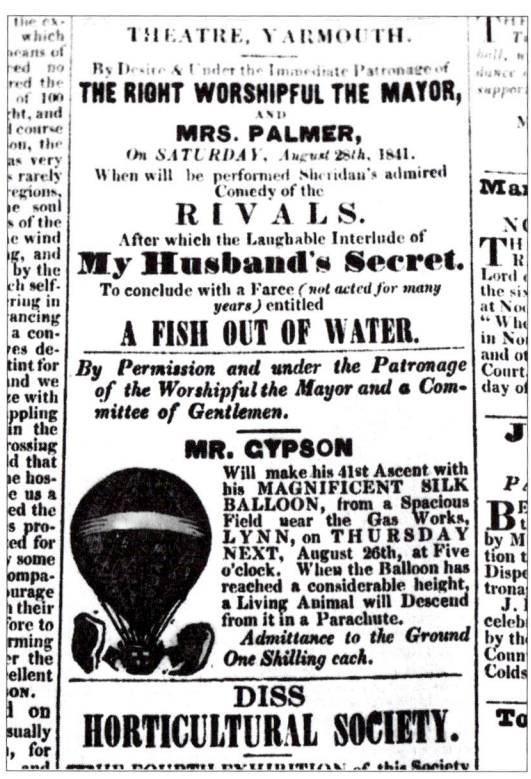

Advertisement for Richard Gypson's balloon exhibition in King's Lynn on 26 August 1841. (Ray Wilson)

Gypson moved on to King's Lynn later that month. The Norwich newspaper makes a perfunctory mention of an ascent from the town on Thursday 26 August, the aeronaut being accompanied on this occasion by Mr Duckworth Nelson, but no destination is mentioned.

Remaining in King's Lynn, he advertised another ascent for 7 September but due to bad weather this was postponed until the 9th, when it had to compete for attention with the annual regatta. The wind was less than helpful that day blowing steadily towards the sea until mid-afternoon. From time to time small test balloons were sent aloft to check its direction and later, high scudding clouds suggested more westerly air currents.

Gypson, anxious no doubt to cover his expenses from his fee-paying passengers, had arranged to take up no less than four people in addition to himself. Mr John Sugars and a young (unnamed) lady together with Mr Withers and Mr Arthur Peek, would be his companions. This was not to be, however, for the balloon had insufficient buoyancy to lift them all. Mr Sugars and his lady friend were persuaded to give up their seats in an effort to relieve the situation.

The balloon, duly lightened, rose sluggishly and crabbing sideways. Its netting became entangled with lofty poles supporting the canvas protecting the arena from prying (and non-paying!) eyes. Once again the balloon was hauled down and on the advice of Mr Gypson it was Mr Withers's turn to vacate his seat.

This time the release of weight was sufficient for the balloon with its two remaining occupants to rise. At first it drifted towards Black Peter sands in the Lynn Channel but managed to pick up an easterly current towards Dersingham and Sandringham.

Advertisement for second ascent by Richard Gypson at King's Lynn on 7 September 1841. (Ray Wilson)

Passing through clouds, Gypson then described his unusual, novel and potentially dangerous method of returning to earth at Dersingham. He claimed to have emptied the lower half of the balloon of gas by releasing the safety valve. Air pressure then caused all the loose, lower silk of the envelope to be forced into the upper cavity of the restraining net to form a canopy and the aeronaut and his companion were brought gently to earth by the parachute-like shape. Gypson claimed this was the third such occasion that he had descended by this method and in view of a passenger being aboard on this occasion, there seems little reason to doubt his word.

From these, Gypson's first exhibitions in the region, gradual changes can be discerned regarding the spectacle of ballooning. While it must be said that Charles Green did not embark on his balloon events without an eye for financial recompense, he seems altogether more subtle and cautious in his approach. There is clear evidence that the new breed of aeronaut was more spectacle-oriented than his predecessors and confirmation of this pattern can be found from reports of the next exhibition, organised by Richard Gypson in Stamford on Friday 1 October 1841.

In those days, it was always the inflation process that drew in wealthier elements of the audience – perhaps it enhanced their feelings of status, privilege and superiority – and as such was the principal source of income for these aeronauts. Gypson's balloon, *Royal Standard*, was no exception to the rule as far as splendour was concerned, being, he claimed, the largest yet seen in Stamford. That was not so but Gypson's showman-like repartee soon glossed over the detail.

Principal among Gypson's three companions for the journey was none other than Octavius Simpson who, it will be recalled, featured in Charles Green's ascent from Stamford in 1826. Obviously a man of adventure, nerve and some financial substance, it comes as no surprise to discover that he went on to become mayor of Stamford in 1866. The other two passengers were Mr Samuel Sharpe from Stamford and Mr Warsop Jr of Alconbury Hill.

Although the balloon was said to be the largest seen thus far, it was not planned to carry four occupants but due to the excellent quality of the gas, this feat was accomplished successfully. When piping gas direct from a gas works to a balloon nearby, it was quite possible, with the co-operation of the gas manager, to draw off a supply of gas with a low specific gravity. Gas given off at the beginning of the distillation process, while best for illumination purposes, was heavy and unsuitable for ballooning. However, gas produced much later in the process burned with an almost invisible flame and was infinitely lighter. By arrangement, and usually for a handsome price premium, – in 1830, for example, 20s (£1) per 1,000cu.ft compared to the domestic price of 5s (25p) – a balloonist could obtain gas with the best lifting potential. If the aeronaut was unable to set up adjacent to a gas works, then by tapping into a town gas main he took a bigger gamble on the lifting power of the gas used.

However, returning to Mr Gypson at Stamford, the inflation process, begun at 6 a.m., was completed at 4.30 p.m. that day. The balloon climbed rapidly to 6,600ft, remaining in view for many minutes before drifting gently northwards. Later, Gypson's passengers recalled:

> slight trepidation, arising from the novelty of our position, was soon exchanged for the most strange but pleasurable sensations, tinged with feelings of excitement and ecstasy, as the earth sank deeper and deeper from us.

In the still air that day, travel through the sky was slow. After forty-five minutes the balloon had progressed only two miles, to the London Road area of Burghley Park. Furthermore, the lifting power of the balloon, already strained by the weight of four occupants, was diminishing rapidly as the day cooled.

Richard Gypson decided it would be prudent to set down in the park itself, where he invited Mr Warsop to alight from the car, thus enabling the balloon to rise once more. At 100ft it was held steady by a group of helpers below holding on to the trail rope. In this manner the balloon was then towed across the park to Burghley House, where the Marquis and Marchioness of Exeter, together with a large crowd, had gathered. Duly inspected by nobility, the balloon was deflated and packed away. The intrepid voyagers were then entertained by Lord Exeter, who on interviewing Mr Gypson, thanked him for the compliment paid in bringing the balloon to Burghley House. His Lordship is also said to have contributed liberally towards defraying the expenses of the ascent and expressed his desire for Mr Gypson to make a second ascent on 7 October under his patronage again.

Later, though, Gypson abandoned the idea as he could not raise sufficient additional financial support among other influential inhabitants. He was also disenchanted with the proposal because the ascent of 1 October had not turned out to be as lucrative as he had anticipated. In what seems to be a blueprint for air shows throughout both centuries, too many people had worked out that they could find good vantage points all over the town (including the roof of St George's church!) and had no need to pay to go into the enclosure.

Disappointed – but not downhearted – Gypson continued his aerial progress round the region and his forty-fifth balloon ascent on 14 October 1841 brought aeronautics to the town of Spalding for the very first time. Although Spalding Gas Works first opened in 1832, its original owner had been less than successful and it changed hands in 1841 when Mr Maples bought the business. This was seen as an opportunity both to celebrate and to advertise his new venture, and what better way than by filling a balloon with gas made by the new owners.

Inflation was carried out on the spacious lawn of Willesby House, home of Mr Brightman, a surgeon, adjoining the gas works on Albion Street. Bad weather dogged preparations but at 5 p.m. Gypson took off alone in a stiff south-westerly breeze. Within ten minutes he was lost to view having entered, in his own words:

> A dense and gloomy cloud in which my position was most uncomfortable from the cold and damp of the vapour floating around me. Even the edge of the car was covered by small globes of water, from condensation of the humid atmosphere. At length I rose into a more rarefied atmosphere, leaving beneath me the strata of clouds through which I had lately passed and which presented a horrible and dismal appearance, resembling huge mountains of dirty ice.

Flying blind in such conditions must have been an unnerving experience, but Gypson was amply rewarded for this discomfort as he drank in the breath-taking view presented him now. Although his journey was intentionally short, he penned a poetic account of what he saw. It should be remembered, too, that although his account might seem flowery to modern eyes, he was witnessing panoramas few had seen before nor would be privileged to witness until the age of the airliner.

> The atmosphere was still gloomy but soon the clouds presented beautiful objects for admiration. Here a young silvery sea of clouds were floating gracefully on the left and on the right others of light blue and vermilion. Poet and philosopher might:
>
> *Gaze and turn away and know not where*
> *Dazzled and drunk with beauty, till the heart*
> *Reeled with its fullness.*
>
> Nature was entirely noiseless up there, even the wind at this elevation is silent. [This crafty fellow is recycling some of his earlier lines for a new audience]
>
> I floated gently along, the lonely stillness being interrupted only by the progress of the car and its colossal ball which, self-propelled, seemed from the waving motion of the loose un-inflated portion of the silk, like a mighty eagle fluttering in the blue ether. Recollecting my promise of an early return to Spalding and knowing that Mr Rainey of the White Hart Inn and the committee were following me in a post-chaise, I re-entered the clouds. When over Holbeach marshes I descended safely. Not being provided with a grappling-iron, though a

deficiency of ascending power in the balloon and, in consequence of the chicken-hearted conduct of an individual below, who could have rendered assistance, I drifted a considerable distance before being finally overtaken by my friends from Spalding.

Mr Smith's Wharf Yard, Sleaford, on 26 October 1841, was the venue for his next voyage but being late in the year the weather was rough and a boisterous wind put paid to any ascent that day. Rising early next day Gypson found the weather had improved a little so the inflation process was started. As the hour approached for the ascent the wind increased in strength, and it was with great difficulty that the partially inflated balloon was prevented from breaking loose from its moorings.

Undeterred, Mr Gypson pressed on with the inflation but at 3 p.m., the allotted time for take-off, heavy rain was falling. Buoyancy was poor and the combined effects of a high wind, humidity and the additional weight imposed by a rain-soaked balloon made it impossible to accomplish anything spectacular. Indeed it was highly probable that a disaster might occur at any moment. Faced with these appalling conditions, postponement seemed an inevitable and cautious step, but Richard Gypson threw caution to that boisterous wind and set off alone. Having committed himself to the expense of filling the balloon and having already postponed the event once, was he perhaps driven on by the financial implications of failure?

Feeling, no doubt, that he had fulfilled his obligation under trying circumstances, the balloon came to earth after just two and a half miles in the parish of Quarrington. Nursing slight bruises from the landing, he announced that, due to the advanced time of year and the consequent poor weather prospects, he would make no more ascents that season, despite many invitations he claimed to have received from several towns in the neighbourhood.

Richard Gypson never returned to the region again, although his ballooning activities continued until 1849, when he was still only thirty-eight years old. During his ten-year career it is believed he made about 100 flights having completed forty-nine of these between 1839 and 1841, the frequency clearly declined rapidly thereafter. This slowdown is unexplained but his most notable trip, on 6 July 1847, almost resulted in his death when the balloon burst. He and two companions – one of whom was Henry Coxwell, of whom we shall hear more later – were saved only by the fabric of the balloon being held within the restraining net, which transformed it into the shape of a parachute. It will be remembered that in 1841, at King's Lynn, he demonstrated how such a feat could be carried out and clearly this experience stood him in very good stead.

Nevertheless, after such a frightening experience, perhaps he chose not to push his luck too far. Perhaps though, he gave up for financial reasons for, sadly, it is recorded he fell on hard times and was imprisoned for theft in May 1853.

Charles Green of course was still going strong at this time but the only balloonist to visit Lincoln since his ascent in 1830 was the relatively unknown Mr Wadman on Monday 17 July 1848. As a result of its rarity this event attracted thousands to the city, including 400 people arriving on a specially laid-on train excursion. 1,500 paid to watch the balloon being filled with coal gas in the grounds of the Horse and Jockey Inn during the day. Then Holmes Common saw an enormous crowd gather to witness the balloon carrying Mr Wadman, accompanied by William Atkinson, take off and drift gently in an easterly direction. Mr Fisher, proprietor of the Durham Ox Inn, wanted to go aloft and took his place in the car but his weight proved too much for the lift capacity of the balloon and he was asked to give up his seat. This journey was unremarkable and lasted for just forty-five minutes. The aeronauts returned safely to earth on Mrs Gamble's farm in Metheringham Fen, fourteen miles from the take-off point.

After a lapse of seven years Mr J.W. Charnock, proprietor of a balloon, announced to the people of Boston that he had arranged 'A Grand Balloon Ascent'. Furthermore, it was announced that this coal gas balloon, named *Victory*, would be 'piloted' by Captain Chambers RN. This is the first use of the term 'pilot' noted in the region. Titles for aeronauts appeared in popular usage, too, and could cause some confusion. Whether self-bestowed, honorary or legitimate, the terms 'professor', 'captain', 'monsieur', 'mademoiselle', and even pseudonyms, were favoured as the century wore on.

It was the first and only, occasion that this 'Captain' Chambers RN visited Boston. However, J.E. Hodgson (1924) has attributed other balloon events held between 1851 and 1854 outside the region, to a certain 'Lieutenant' Chambers RN. It is recorded that Chambers had a son, also a balloonist who was killed in a balloon accident in Nottingham in 1863, but the *Annual Register* magazine of 1863 reported the death as that of a 'Mr' Chambers. In view of the dates mentioned and the confusing use of different titles it is unclear precisely as to which gentleman was being described at the Boston event of 1855. Perhaps an explanation is simply that Lieutenant Chambers gained promotion, by one means or another, and actually continued his aeronautical career into 1855. Such are the delights of research!

Great numbers of people assembled in Dawson's Yard in Wide Bargate, Boston, on Wednesday 1 August 1855 to witness the ascent. Vast crowds jostled for a view along Bargate and Horncastle Road, while every available vantage point on the tops of houses in the vicinity was occupied. At 6 p.m., the appointed hour for take-off, the balloon was not yet half inflated. Even though the local brass band played valiantly to relieve the tedium of slow inflation, it was soon evident to all that the balloon had no chance of ascending that night. Mr Charnock, the owner, was most apologetic, taking great care to point out that it was the fault of the gas company in not laying a sufficiently large service pipe, a matter destined to become a bone of contention at many events. In compensation for the great disappointment (and no doubt his reluctance to hand back the takings) he would allow it to ascend the next evening from Bargate Green without any further charge being made. With his promise echoing in their ears the crowd, although grumbling loudly, dispersed in an orderly and peaceable manner.

Rescheduling the event, however, proved to be a source of friction between Mr Charnock and the mayor of Boston. Next day Charnock sought from the mayor the necessary permission to use Bargate Green. Even though authority to grant such permission was quite within the power of the mayor, he responded evasively to the request and referred Mr Charnock to the lessees. Nonplussed, Charnock took the mayor's answer as tantamount to a refusal, took 'the hump' and prepared to leave town. On hearing of this, the mayor hastily reconsidered the matter and later that day sent for Mr Charnock, granting him permission for the event to proceed.

Better arrangements were made with the gas company this time and the ascent was scheduled for the evening of Friday 3 August 1855. Supplied by a much larger gas pipe, filling took a mere three hours and at 7 p.m. the balloon rose majestically into the clear evening sky

Captain Chambers carried no passengers with him and his flight was but twenty minutes in duration. With the wind from the southwest he considered it prudent to descend quickly to avoid being blown out over The Wash. He came down safely on Mr Seymour's Farm at Wrangle Tofts, about half a mile from the sea bank and ten miles from Boston.

Despite the coming of the showman era there were still those who maintained a more idealistic approach to aeronautics. Following the retirement of Charles Green, the mantle of leading exponent of aerostation fell upon Henry Coxwell, whose career as a pilot began in 1848 before turning professional in 1852. Coxwell fell in love with the idea of ballooning when he attended a Charles Green event at Rochester in 1828. He became an apprentice dentist in 1836 and did not manage to get into the air until 1844, when he went aloft as a passenger with George Gape and John Hampton.

Henry Coxwell was an advocate of using balloons for military and scientific purposes, and it is in pursuit of the latter that he made an air journey to Rutland in July 1862 accompanied by James Glaisher FRS, on the first of Glaisher's three scientific flights made with Coxwell that year. Glaisher was an English scientist specialising in meteorology as superintendent of the Magnetic and Meteorological department of the Royal Observatory in Greenwich. He, too, recognised the potential of the balloon as a tool to facilitate high altitude scientific observations and he was dismissive of any other role for free ballooning,

The aeronaut Henry Coxwell, who collaborated with James Glaisher on scientific balloon flights during the 1860s. (Via Brian Cocks)

such as for exhibitionism or travel. In 1866 Glaisher became one of the founders of the (Royal) Aeronautical Society and was its treasurer then chairman for a time.

He persuaded the British Association to fund a series of daytime and nighttime scientific flights that would be spread over many years. He engaged Henry Coxwell who, in the absence of a suitable existing balloon, manufactured a new one and managed the ascents with himself on board as the observer. This was not Glaisher's original intention but as he said later:

> Notwithstanding all these difficulties [over finding a balloon] and the efforts I had been obliged to make to overcome them, I found that in spite of myself I was pledged both in the eyes of the public and the British Association to produce some results in return for the money expended. I therefore offered to make the observations myself.
>
> I accustomed myself to the use and manipulation of the instruments in a limited space and considered how best to group them on a board to serve as a table in the car of the great balloon.
>
> The experience that I have acquired … will show how much those [people] are in error who think that observations in the higher regions can be made well enough by the first person that comes along. The novelty of the situation, the rapidity with which all the observations must be made and the smallness of the space require that the observer should have previously had considerable practice in the use of the instruments under all circumstances.

On 30 June 1862 came the time to put all the theory to the test. Coxwell and Glaisher took the brand new 90,000cu.ft capacity balloon to Wolverhampton gas works for the first flight. It was made not of silk but of 'American' cloth, which was a strong material and as yet did not have a name but several years later Coxwell christened it – not surprisingly – *Mammoth*. The reason for choosing Wolverhampton was that if the wind – which in summer blew generally from the south-west – carried them any distance before they could return to earth, the best prospect of a long run over relatively benign countryside before reaching the sea would be across the East Midlands.

A gusty wind made inflation difficult and disaster struck when the force of the wind bounced the balloon around so much that it split vertically and then horizontally around a weak seam. All the gas was lost and operations were abandoned while repairs were made over the next two weeks. During that time the gas company, much to the delight of the aeronauts, agreed to siphon off regular amounts of 'light' gas from the distillation cycle and store it in a separate gas holder ready for use when the balloon was repaired. This most agreeable gesture by the gas company made a particularly valuable contribution to the achievement of such great altitudes during Glaisher's ascents from that town.

Dogged by more bad weather, it was not until 17 July that the pair felt they had no option but to 'go for broke' and begin the experimental flight programme. So, despite the still alarmingly high west-south-west wind, Henry released the balloon at 9.43 a.m. with the barometer reading 29.50in and the temperature 55°F. James recalled his feelings:

> The movements of the balloon were so great and so rapid that it was impossible to fix a single [scientific] instrument in position before quitting the earth and this state of affairs was by no means cheering to a novice who had never before put his foot in the car of a balloon.

In six minutes they reached the first cloud layer at 4,500ft. On upwards at quite a pace they passed through more cloud with Glaisher now working feverishly at his observations. It must have been a mixture of his hard work and no doubt some apprehension that caused his pulse rate to shoot up from seventy-six beats per minute on take off to nearly 100 half an hour later, while Coxwell's had risen from seventy-five to eighty-six. At 20,000ft they were impressed by the way the pale light blue of the sky at ground level had changed to an 'intensely deep Prussian blue'. Now their world was silent and every sound in the car magnified:

> Palpitation of the heart was very susceptible, so much so that each man could hear the beating in the breast of the other. The ticking of [Glaisher's] watch was remarkably loud… and the rustling when turning over the leaves of [my] notebook represented the rushing of a high wind. The air was very dry and our hands and lips were blue but not our faces. The temperature was 42°F and began to decrease with wonderful rapidity.

After an hour and a half in the air, at 11.10 a.m., the balloon reached its highest altitude (about 26,000ft) when the barometer read 11in and the temperature had fallen to 16°F (or 16° of frost). Both men were feeling the effects of altitude, lack of oxygen and the cold. Glaisher recalled:

> Breathing, which was observed to be interfered with when heart palpitation commenced, again became affected and the cold was felt. Henry Coxwell only found it necessary to throw on one additional coat and I wrapped a cloak around [me] for a short time. At 10.57am [I] had the feeling of sea-sickness with its uncomfortable manifestations and again at 11.07am but it was not so prolonged.

At 11.25 a.m. Coxwell became uneasy as to their position and discerning The Wash in the distance, deemed it prudent to begin a rapid descent. The balloon was seen coming down by people below and one of the local gentry, Mr E.G. Baker, filled with foreboding about the rapidity of the descent:

> prepared restoratives and gave such orders as he feared would be called for on account of what he expected would be the personal injuries which the voyagers would suffer.

The touchdown point was on the parish boundary of Langham and Ashwell, near Oakham in Rutland, having flown a distance of eighty miles. Mr Baker's concern was, thankfully, unfounded for though the instruments were badly smashed, Glaisher sustained only a slight injury to his face and hand, walking away with a headache while Coxwell emerged unscathed.

When the balloon touched down the first person to rush up was a local yokel. On being asked by these men from the sky where they were, he replied in the vernacular, 'near Oak'am, zur!' The aeronauts took that to mean they had come down near Holkham, Norfolk, and the man was asked to send a messenger immediately to Lord Leicester's house and tell him that Mr Glaisher, to whom he was known, had arrived. It took a while to sort that one out!

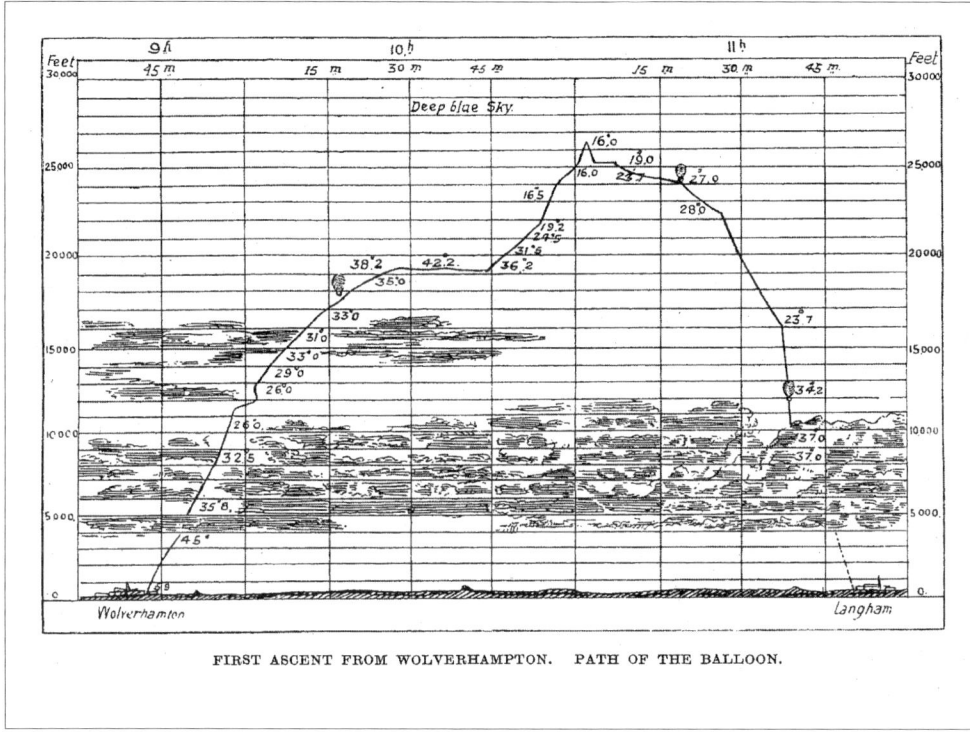

FIRST ASCENT FROM WOLVERHAMPTON. PATH OF THE BALLOON.

Above and below: The flight profile of Coxwell and Glaisher's first aerial scientific expedition on 17 July 1862, when they flew from Wolverhampton to Langham in Rutland. (Brian Cocks)

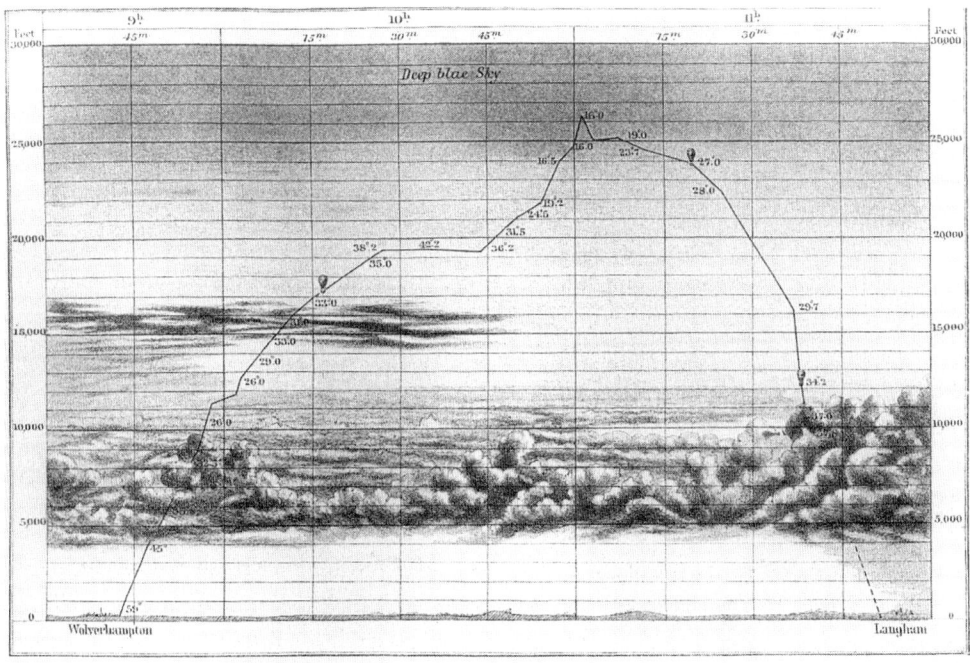

Path of the Balloon in its ascent from Wolverhampton to Langham.
July 17th 1862.

Henry Coxwell and James Glaisher had made a good start to their programme of scientific flights but the next one, in September that year, was very nearly 'curtains' for them both. That flight, however, did not bring them into the region covered by this story, so we shall have to leave them in Wolverhampton, planning the journey that would make them truly famous.

In stark contrast, although with archetypal Victorian gentility, Stamford Horticultural Society began to hold an annual Grand Fête and Rose Show, in the grounds opposite Rutland Terrace. Hundreds of visitors were attracted to the show by the announcement that Mr Adams Jr, from the famous Cremorne Gardens, London, would make an ascent in his balloon. Bright and early on the morning of Wednesday 8 July 1863, the inflation process began but the affair was to end ignominiously for Mr Adams.

Ominously, from the outset the balloon seemed reluctant to take shape. By the scheduled time of departure for special trains that had brought many visitors to the show, there was still insufficient gas in the balloon to justify even the slightest hope that those excursionists would be able to see the ascent. Doubts as to whether Mr Adams would ever get into the air were confirmed finally at 8 p.m. when the attempt was abandoned. Once again the blame was laid at the door of the local gas company, when Mr Adams explained that with a pipe of only a 3in bore and starting at 8 a.m., it was impossible to pass the required volume of gas in the intervening twelve hours. On the other hand, the gas works manager countered this accusation by asserting that 3,000cu.ft of gas in excess of the volume requested had in fact been pumped that day. There was no doubt in the minds of those present that a considerable leakage had taken place because the air around the showground reeked with the stench of gas all day.

Having made the decision to abandon the ascent, Mr Adams discharged the gas that had been collected, packed up his equipment and had it removed to the railway station. That such copious quantities of inflammable gas could be allowed to waft around a public event in so careless a manner, risking a calamitous explosion or fire, would not bear thinking about in modern times.

Naturally the cancellation caused great consternation among the crowd, particularly in the light of a rumour circulating that Mr Adams had disappointed people of another town in a similar way. The crowd grew ugly and he was hooted, jeered and jostled all the way to the station with the police having some difficulty in getting him away safely.

By the 1860s ballooning had almost taken on the appearance of a circus 'turn'. Earlier associations with the proprietors of prestigious, opulent 'gardens', such as London's Vauxhall and Cremorne, Birmingham's Vauxhall and Manchester's Belle Vue were over, as those gardens themselves became unprofitable and gradually fell into bankruptcy. From being an inevitable star attraction at fenland venues, the declining popularity of balloons could be detected from the way in which it was pushed down the order of events, as shown by several advertisements of the time.

As a spectacle, though, the sight of one of these monsters at a fête or gala, could still thrill a crowd but somehow they lacked the pulling power to achieve this entirely on their own. Like all such innovations, the second half of the nineteenth century saw an upsurge of balloonists, all vying for the public's attention in an effort to cash in on its novelty value. This euphoria, however, was wearing thin. Crowds wanted more sustained entertainment value for their money – to the extent, even, of transposing Vaudeville-style programmes of entertainment to outdoor occasions. No longer were they content merely to pay for the privilege of seeing balloons inflated or take off, or for that matter, to watch without paying.

It was just not exciting enough any more.

This, then, was the scenario in which Emanuel Jackson of Derby conducted his ballooning exhibitions which took him to Stamford in 1869. Billed as the 'Midlands Aeronaut with his Monster Balloon', Jackson was part of a circus-like programme assembled for the Great Star Gala of Monday 6 September 1869. Fêtes, galas, country and horticultural shows were much in evidence now across the land. People of all ages and from all walks of life could enjoy several days out during a year, on the recently introduced bank holidays and other festive occasions.

Heavy clouds hovered over the field and drenching showers fell at frequent intervals both during the morning and afternoon. Performances by the cast of entertainers were halted until the evening. In consequence, there was a poor attendance and it was believed the organiser's expenses would scarcely be met.

Moreover, it is possible to judge the degree of apathy towards ballooning, from the rather terse wording in the previously enthusiastic *Stamford Mercury*. A mere five words, 'including a fire-balloon [sic] ascent', was all Mr Jackson's efforts rated and no detail about his ascent was forthcoming. From the use of the unusual term 'fire-balloon', it may be deduced that Jackson's balloon was of the hot-air type rather than the more usual coal gas model. This opinion is supported by the venue chosen, which was some distance from the gas works, and from the apparently unimpressive and therefore possibly quite short duration of the flight. Further evidence to support this view can be found in a report of an event in Stamford some five years later when, in honour of the wedding of the Duke of Edinburgh, another gala was held on Wednesday 2 September 1874.

Once again the venue was a three-field site on Ryhall Road, opposite the infirmary. Here, blessed with fine weather this time, 'Sangers Monster Fête' drew a crowd of 14,000 people from the town and surrounding countryside. Proclaiming the event to its readers, the *Mercury* announced that a 'real balloon … the largest ever manufactured in this or any other country…' would be a star attraction. Significantly though, it was stated clearly that this balloon would ascend from the gas works and not the field in which the fête was held.

Captain B. Metcalf and Doctor Ross were the intrepid aeronauts for this exhibition. Lifting off with ease they drifted over Ryhall Road near the fête ground, before taking a course to the north-east. No final destination was recorded but Captain Metcalf had stated earlier that it was his intention to make only a short trip.

Victorian Boston was an expanding maritime town whose council intended to display its grandeur to the outside world at every opportunity. Just such an opportunity occurred in September 1879 when the newly built Fydell Crescent was opened with a grand gala whose star attraction was a coal gas balloon ascent from the new street itself.

Billed as one of the most celebrated English aeronauts of his day, James A. Whelan of Huddersfield was engaged to make an ascent in his balloon *Excelsior* (formerly named *Duke of Edinburgh*, according to the event advertisement). Several Boston gentlemen expressed a desire to go aloft with Whelan but none was willing to pay his asking price

Part of a poster for Stamford Gala on 6 September 1869 at which Emanuel Jackson of Derby made a balloon flight. (Stamford Museum)

so they missed out to a Mr Beatson, an acquaintance of Whelan's from Huddersfield who, prepared to pay the fee, rushed down by train to make the ascent. It seems amazing – even when viewed from the twenty-first century – that in those days word could be sent (probably by telegram) to Huddersfield and Mr Beatson could still reach Boston by train before 6 p.m. the same day.

Supervised by Mr W. Stout, inflation of the 25,000cu.ft *Excelsior* was organised by the Boston Gas Light & Coke Co., whose directors gave the gas free of charge as their contribution to the festivities, much to the delight and surprise of Mr Whelan.

At 6 p.m., amid loud cheering, on the command of 'Let go, all!' twelve strong men keeping the bucking balloon on the ground released their hold on the ropes and *Excelsior* rose majestically on the brisk south-west wind. Interviewed later Mr Whelan described the voyage:

Not being a swimmer I watched the wind direction carefully during the inflation, studying my map and pocket compass frequently, as I had no desire to go to sea. As we left the ground, the decorations in Fydell Crescent looked very pretty and there was a dense mass of people.

Our greatest height was 3,000 feet as we swept north east at a rapid rate. Just beyond the town limits [High Ferry, Sibsey] we encountered a cross current that swayed the balloon roughly and we now appeared to drift towards the sea [The Wash].

As landing in the dark is not an agreeable thing, we decided to come down as quickly as regard for our necks would allow. Aiming for a suitable field ahead, I opened the gas valve but our speed swept us a mile past it before the grapple made a firm bite. Due to the strong wind our landing was very bumpy and with some bruising we came down at Freiston Ings in a field owned by Mr James Needham. Our 'roughing up' fully justified my declining to have a third person with us, as an inexperienced person would have been pitched out of the car and taken home in a condition in which his wife would hardly have recognised him!

James Whelan described his flight as 'our voyage' but that was much too grand a term. After all, his flight only lasted fifteen minutes, the highest altitude reached was 3,000ft and they were back in Boston by 9 p.m. A spectator watching from the top of Boston Stump summed it all up rather tartly, 'The voyage of the balloonists had been like the gallop of the proverbial donkey – short and sweet! Of course it is a pure assumption to imagine that it was sweet seeing that I have never tried it.'

But fashions in public entertainment come and go, and it is not at all unusual to find that, after a period of years, public interest in galas and fêtes also began to wane. This is not a problem identifiable with any particular period of history but rather a reflection of the public's fickle taste. It was one thing to attract an audience but quite another to keep its interest alive year after year. The public was ready for a change and now was the time for the world of ballooning to undergo another change of direction.

Since 1873 Spalding Cricket and Athletic Club's annual Sports and Horse Show had become the principal and most prestigious event in the fenland region. Competitors from all corners of the land were attracted by its reputation but, by 1890, both spectator and competitor attendance at this show had been in decline for a number of years.

In an editorial comment following the 1888 show the *Lincolnshire Free Press*, reflecting the mood of the times, was moved to observe: ' … in our opinion a more spirited and less stereotyped programme would have to be adopted if the event was to be continued'. What was to be done, then, to keep the show alive?

In an imaginative, if unadventurous, move the organising committee engaged a team of championship trick cyclists to perform at the 1889 event and succeeded in doubling the gate receipts. Then, in 1890, the committee really excelled itself by entering the world of aeronautics for a spectacle worthy of the name. This first attempt would be memorable but for vastly different reasons than those envisaged by the organisers!

It had long since passed from general memory that parachuting had been a source of excitement to participants and watchers alike at ballooning exhibitions since the beginning of the nineteenth century. Parachuting had been put to little practical use during the intervening years and indeed had passed out of favour following the tragic and highly publicised death of Robert Cocking in 1837, who, coincidentally, had been carried aloft for his ill-fated experiment by none other than Charles Green.

It was not until mid-1888 that parachuting returned to the public eye when an American, 'Professor' Thomas Baldwin, brought his daring parachute act to the Alexandra

Palace, England. Great controversy and alarm accompanied his arrival. Questions were even asked in the House of Lords about the wisdom of permitting such 'dangerous and demoralizing' exhibitions. Nevertheless, Baldwin went ahead with his parachuting tour and won over the public, who relished the drama of the jump and – one suspects – the possibility of a mishap. Nothing changes, even a century later!

Having re-established parachuting from beneath a balloon as a money-spinning spectacle, it was not long before Thomas Baldwin had a string of imitators. One of these aeronautical adventurers was 'Professor' Russett from Nottingham who, claiming to be the tender age of nineteen, billed himself as 'the youngest aeronaut in the land'. Described by the *Lincolnshire Free Press* as being 'of gentlemanly bearing, intelligent and exuding coolness and presence of mind', he also claimed to have made numerous parachute descents during the past year.

August bank holiday Monday 1890 dawned and Professor Russett, accompanied by his wife and his manager, duly arrived in Spalding on the day of the show. The plan was for Russett to ascend from the cricket field at the rear of the Black Swan Inn (the site of the town football ground in 2007). He would then make a parachute descent into the arena.

The balloon was of conventional appearance, having a capacity of 15,000cu.ft to be filled, in this instance, with coal gas. This task was accomplished during the afternoon of the show by pumping gas through a specially laid 6in-diameter pipe connecting with the town gas main in nearby New Road. Once inflated the balloon was held down by ropes and sandbags until the crucial moment of launching.

The parachutes used followed the basic 'limp canopy' design in service in modern times but the main difference being the absence of a backpack container for the canopy. The canopy was suspended in the manner of a rolled umbrella from the balloon restraining net or the seat ropes and secured to a primitive body harness worn by the jumper. Some parachutes – for there were no standard designs – had provision for a stirrup-like loop in which the jumper could place a foot to help balance. The whole contraption was secured to the balloon net or ropes by a fragile ribbon, which broke easily with the parachutist's weight as he or she dropped away from the balloon.

Pieces of sheet iron were sewn into the fabric at the top of the balloon but slightly off centre. This provided a means of recovering the sphere after the aeronaut had jumped. The balloon would only remain in an upright attitude by the aeronaut's weight acting as a pendulum. Consequently when the person jumped, the balloon turned over, the gas escaped from the neck of the sphere and it floated down to earth – hopefully without causing damage to life or property!

This was, then, the general arrangement for exhibitions involving one or more parachutists, all of whom would jump from beneath the balloon. Again designs varied but jumpers would either sit on a trapeze-like bar from which they would drop off or they might grip and hang from such a bar and simply let go at the desired altitude. Later it will be seen that multiple jumps would be made from a conventional basket suspended beneath a balloon, manned by a 'pilot' who would remain with the craft.

Returning now to that gay Victorian bank holiday scene, the crowd waited for Professor Russett to present his daredevil act. It waited in vain!

The committee's good intentions and the town's introduction to the spectacle of parachuting were marred that day by events which degenerated into a cross between a

Whitehall farce and a soccer riot. At the centre of the storm was the age-old issue of value for money. Entrance fee to the show was doubled this year to 1s (5p) and spectators were persuaded to part with their hard-earned wages on the expectation of witnessing a unique spectacle. Over 7,000 people, including 4,000 arriving by special rail excursions, flocked to the sports field, on the tide of a bank holiday carnival atmosphere – so many that the young professor declined to take off from the main arena, fearing that both he and the crowd would be at risk. Instead, he arranged to lift off from the adjacent field in which the balloon had been inflated before being moved to the arena.

Announcing this change, together with the non-appearance of other star performers at the sports, unsettled the crowd but it was a small worry compared to the discovery that the balloon, supposedly filled by now with 15,000cu.ft of gas, had insufficient buoyancy to lift off! Despite a further 2,000cu.ft being pumped into the balloon – now in danger of bursting – it obstinately refused to show the slightest sign of being able to carry Professor Russett aloft.

Afternoon wore on into evening. By 7 p.m., with the crowd growing restless, it was decided that as there was just sufficient buoyancy, the balloon would be moved back into the arena as a goodwill gesture. It was then that disaster struck.

Being unstable, the sphere turned completely over as it was being manhandled, thus allowing the gas to escape. Soon it became obvious to the already impatient watchers that this signalled the end of the spectacle and all hell broke loose! With the arena echoing to hoots of derision the professor was loudly condemned as a sham and a fraud. He was assaulted and the balloon trampled upon by the mob, oblivious to the very real danger of explosion and fire from cigarettes and pipes igniting the escaping gas!

Russett and his team escaped serious injury only by running for the safety of the cricket pavilion and barricading themselves inside. They were imprisoned for two hours before escape became possible and only then after Russett had changed clothes with a certain Mr Cooper, who bravely climbed out of a rear window and drew off the mob in the gathering dusk. The frightened party succeeded in escaping to a safe house in another part of town, where they no doubt reflected at length upon their ordeal.

Reporting these farcical events the *Spalding Guardian* quoted Professor Russett's explanation for the fiasco as, 'in a small place like Spalding, the gas company would use low grade coal which would produce poor quality gas with inadequate lifting powers.' Called to account before the organising committee, he repeated his explanation only to have it vehemently rejected by the gas works manager, Mr Woodward, as 'utter nonsense!' According to Mr Woodward only 4,000cu.ft of gas had been pumped to the whole town on the day in question. It was his belief that Professor Russett and his helpers had allowed atmospheric air to enter the balloon during filling. This later, he insisted, would negate the lifting power of the gas.

Following the 1890 debacle, for the next few years the show committee, perhaps wisely in the circumstances, chose a series of more down to earth attractions, such as trick cyclists and high wire acts, but attendances at the show declined once more. Thus it was 1894 before the *Spalding Guardian* heralded another 'visit to cloudland' as the main attraction of that year's Sports Festival, on Monday 6 August. This time, the services of an aeronaut and part-time actor, billed as Professor Charles Baldwin of Farnham, Surrey, were engaged. He arrived in Spalding accompanied by his wife and his manager, Captain J. Seymour-Cooke, designer of the parachutes to be used.

In the pre-show publicity material it was claimed that the twenty-four-year-old Baldwin had been parachuting for five years and this was to be his 130th ascent and descent. A small pen picture of this intrepid entertainer was found in a contemporary magazine called *The Era*.

> Mr Charles Baldwin is a many-sided man, for example, after making a successful [parachute] descent at Portsmouth on Wednesday week [July 1894], he travelled to Worthing and opened in Mr Arthur Lloyd's play *Our Party*, in a light comedy part. On Bank Holiday [August 6] he is engaged to make a parachute descent at Spalding, after which he will journey to Liverpool to repaint the scenery of the play *Ballyvogan* and then stage manage the piece, in which he plays a principal part.

What a fellow! Most definitely a man of many talents and considerable energy!

Bearing in mind the last fiasco, what would the crowd have thought had they known that this young 'Professor Charles Baldwin' was, in fact, not the 'original' article? The original 'Baldwin' refers to an American, Thomas S. Baldwin, a former trapeze and high-wire performer turned parachutist. Tom Baldwin was forty years old when, as mentioned earlier, in the face of great opposition, he re-established parachuting in Britain in 1888, with himself as its indisputable master. He made a great deal of money from his daredevil exploits and it is reported he returned to America in 1890, where he concentrated on balloon construction and promoted his travelling show. The aftermath of his departure spawned a succession of parachuting 'Baldwins' who were no relation to Thomas, but the name itself was synonymous with parachuting success and therefore no doubt deemed good for business.

Happily, though, the 1894 event was subject neither to mishaps, delays nor riots, and on that glorious August Bank Holiday afternoon, before a hushed crowd, Baldwin sailed majestically into the sky above Spalding. At 4,000ft, swaying gently towards the river Welland, he jumped from the seat and floated, gossamer-like, beneath his parachute to earth in Willow Walk near Commercial Road, less than a half mile from the start point. The balloon came down a mile further on, near Fulney House on the Holbeach Road. Upon his triumphant return to the arena, having scrounged a lift in a pony and trap, the crowd went wild, cheering and applauding the intrepid bird-man, while the town band played a rousing version of *See the conquering hero comes*. This, then, was the euphoric climax to the first parachute drop ever seen in the town. Charles Baldwin was the man of the moment and he was invited to return the following year.

Peterborough was a burgeoning city, growing up with the railway age but also a prominent centre for the agricultural area encircling it near the western edge of the fens. It had a long tradition of summer fêtes and galas and joined the popular trend in exposing its population to the wonders of aerostation.

From daybreak in the city on the morning of Monday 6 August 1894, there was a persistent downpour but it cleared before the annual Peterborough gala was declared open. Despite intermittent rain, 7,000 people attended, teeming in by rail excursions as well as from the immediate area. With a balloon billed as the major attraction, other items on the entertainment programme included acrobats, trapeze artists and a magician, in much the same formula as seen in other towns. Under the supervision of that famous aeronautical company, Messrs Green, Spencer & Sons of London it was announced that the balloon would be used both for tethered ascents and an 'away flight' to Colchester if conditions permitted.

Tethered balloon ascents began at 2 p.m. with five passengers being taken aloft at a time. As can be imagined, such a load together with an energetic balloon, called for a substantial mechanism to raise and lower. For this purpose a powerful steam traction engine and windlass was used and it was not long before the engine driver was the weary butt of the inevitable joke, 'mind you hold on to that engine so it's not taken up as well!'

On the day itself, the aerial voyagers into cloudland were charged half a crown (12½p) each for a captive ascent, while for the 'right-away' trip £2 was the asking price. In the event, the promise of a trip to Colchester proved wildly optimistic since the balloon and passengers, with aeronaut Joseph Simmons in command, only reached the village of Coates, a mere seven miles from Peterborough. Villagers, who helped to deflate the craft and pack it up, probably regarded it as manna from Heaven, for it is recorded they were 'liberally regaled at the Bottle and Glass public house by the aeronaut'.

Victorian elegance. A well-dressed crowd watching Green, Spencer & Sons' balloon at Woodston recreational ground, Peterborough on 6 August 1894. The parachutist is believed to be Mr Joseph Simmons.

The next year, assisted once again by Captain Seymour-Cooke, Professor Baldwin attracted 6,000 spectators to the 1895 Spalding August Bank Holiday Show. Each paid 1s (5p) thereby increasing gate receipts that year to £300.

Before the balloon had drifted no more than a half mile downwind, Baldwin gained sufficient height and made his jump. He landed safely in Mr J.T. White's bulb field near the town cemetery and the balloon drifted for two more miles before coming to earth close to Pinchbeck church. Feted by the adoring crowd Baldwin, and his team were guests of honour at an evening garden party held in the grounds of Ayscoughfee Hall. The imaginative committee had clearly hit upon a winning formula.

Euphoria, however, soon turned to horror, as tragedy struck not twenty miles away in Fletton, Peterborough. On that fateful bank holiday in 1895, 10,000 people gazed skywards at the sight of the 20,000cu.ft capacity balloon, *Victoria*, waving gently in a stiff breeze above Fletton recreational ground. That same 10,000 were shortly to witness the death of thirty-six-year-old parachutist, Adelaide Bassett.

Billed as 'Mademoiselle' Adelaide Bassett, she and her companion, 'Captain' Arthur Orton, were professional aeronauts of some experience, having made thirty parachute descents together prior to this event. They also claimed to offer the only double-parachute spectacle in the world at this time. Captain Orton, forty-three, the owner of the balloon, supervised its inflation from a temporary extension pipe laid from the village coal gas main.

Mademoiselle Bassett was, in fact, married but lived apart from her husband, sharing a house with Arthur Orton and his wife in London's East End. She had been parachuting for six years under the tutelage of Captain Orton, having been, she said, 'inspired to take up the aerial profession by the exploits of the Old Professor Baldwin' (an obvious reference to the American Tom Baldwin, who gave many performances in London from 1888 onwards). She exuded confidence, saying 'Why, bless me, it's as easy, once you know how to do it, as walking!' The reporter, clearly infatuated with Mlle Bassett, wrote glowingly, 'She was a fine woman … of rare physique … having auburn hair, plump face, fresh colour and with sparkling eyes and a flow of pleasant talk; a taking woman.' An Amazon, no less!

Captain Orton, for his part, took pains to point out that this was not a 'car' balloon, explaining that journeys were not made in it. It was merely the means by which height was gained for parachuting. Nor was it made from silk in the manner of some of its predecessors. This particular balloon was constructed from prepared Cambric, a fine linen fabric treated on the inside to render it impervious to the gas, much less expensive than silk, although somewhat heavier. In place of a 'car' was a simple wooden seat, just wide enough for two people to sit side by side and it was on this that the intrepid pair took to the sky. At a signal from Orton they would jump from the seat simultaneously. Their falling weight would release each parachute from the balloon netting and both should float gracefully to earth. That, at least, was the plan.

Amid shouts of 'Huzza!' *Victoria* was released by her twelve strong handlers but, made wet by drizzle earlier in the day, she rose sluggishly. Caught by the strong wind, the balloon crabbed sideways toward a tree some thirty yards distant. Wallowing like a porpoise, its netting dragged at the branches, *Victoria* swung into contact with nearby telegraph wires – and tragedy struck.

Mlle Bassett's parachute was wrenched from the netting by the impact with the wires. In an instant, Captain Orton threw his arm across his companion's body to save her from being dragged from her seat. As the balloon freed itself, rising slowly at last, he saw the canopy was billowing inverted but clear of the wires and so drew back his arm.

At the inquest later he recalled saying 'For goodness sake, Addie, don't go!' Whether the young woman misheard him or misunderstood his intention will never be known, but she leaped from her seat at a height of 200ft. Perhaps she thought her weight would have sufficient effect to allow her to fall past the canopy and turn it to its correct attitude before she reached the ground. Indeed, Captain Orton, speaking at the inquest, believed this was quite possible to achieve with sufficient altitude.

However, sufficient altitude there was not and poor Adelaide Bassett plummeted to the ground where she was killed instantly by the impact. Horrified, Arthur Orton clung to his seat until he judged it the bare minimum for a safe descent, then he, too, parachuted down, rushing to the broken body of his companion. A hush fell over the stunned crowd, until the death was confirmed, whereupon the fête was halted and the stunned spectators went home.

Over the following days, news of her death spread to all corners of the land, with most of the major newspapers carrying scathing attacks on the futility and senselessness of such parachuting events. Among these, the *London Evening Standard* was most outspoken, pronouncing:

> No single redeeming feature is attached to these feats of foolhardiness. An idle crowd unfortunately is always to be found ready to encourage these exhibitions. Perhaps the manifest danger to the performer brings out that Tartar instinct which is innate with the Briton as well as the Russian. The law is not bound to tolerate these outbreaks of latent barbarism.

The inquest jury's rider to a verdict of 'accidental death' was 'since no useful purpose is served by these senseless exhibitions, at which the lives of performers are risked, they should be made illegal.'

Despite the pontificating of the press, it seems the public simply got what many lusted for and in that fact, times have changed little since the days of gladiators, knights, wing-walkers, crazy flyers or racing car drivers who draw crowds looking for a thrill from those people who have the guts to provide them with one. Overcome with grief, Arthur Orton vowed never to undertake another parachute drop, but the public outcry seems to have had little lasting effect elsewhere and in Spalding, for example, there was very nearly a repetition of this accident just a few years later.

Seeing a balloonist in action was one of the best reasons for getting drunk! So claimed Samuel Davis, a farm labourer, when summoned before the Wisbech magistrate for being drunk and disorderly the day after Charles Baldwin's next visit to the region. Replying to the charge, the defendant said meekly 'Sir, I was only merry. I went to the sports and got a little excited seeing that man go up in the balloon.' The magistrate, however, was unimpressed and fined Davis 5s (25p). The sports to which the hapless Davis referred were the Wisbech and District Bicycle Club Races, held in the evening of Wednesday 24 June 1896.

Announcement of the Wisbech carnival in 1896 that included a balloon event by 'Professor' Charles Baldwin, billed as the 'Prince of Parachutists'. (*Wisbech Standard*)

Billed as the 'Prince of Parachutists' and ably assisted once again by Captain Seymour-Cooke, Charles Baldwin was engaged to entertain a crowd that grew to over 5,000 people. At 7 p.m. precisely, Baldwin rose into the air beneath his coal gas balloon to a height of 1,500ft.

In view of the outcry following the Peterborough fatality the previous year, it is ironic to note that a press report describing Baldwin's ascent used the phrase 'The leap for life … was taken when the balloon appeared to be almost over the Great Eastern railway station'. His descent took but a few minutes; then it was all over. Perhaps it is a measure of the increasing lack of apparent danger coupled with the improving safety and reliability of parachutes, that the Wisbech event warranted only 200 words within a vast and detailed description of the bicycle races.

Charles Baldwin's final foray into East Midland skies took place at Spalding on the bank holiday of 2 August 1897. For this, his third exhibition in the town, he was accompanied by Captain Seymour-Cooke, who supervised the inflation in Mr P. White's field, at the rear of the Black Swan Inn. With the assistance of Mr Hawkins, the Spalding gas works manager, the balloon, of 30,000cu.ft capacity, took four hours to fill through a temporary 6in extension pipe laid from the gas main in nearby New Road.

Although history is fortunate in that the build-up was prolifically photographed, taking off at 5 p.m. the balloon was actually in the air for little more than ten minutes, half of which were spent in climbing to an altitude of 6,000ft. Baldwin's parachute brought him down safely into the back gardens of the 'Twenty Row' cottages in Winsover Road, while the balloon flopped to earth on the prominent Midland railway embankment about a mile from the show ground.

Many of the 6,000 paying spectators must have wondered what all the fuss was about for so short an item. With this in mind, the promoters realised they needed to look for something a little different to rekindle the public's flagging interest. Balloons, as such, were still monster machines whose sheer size would always fascinate the public but a five-minute parachute drop, on the other hand, had become just another routine event in the long sports and show programme.

Charles Baldwin's balloon and netting is laid out on the ground ready for the inflation process to begin in Spalding, 2 August 1897. (Spalding Gentlemen's Society)

Inflation begins as coal gas is fed into the balloon envelope through a large pipe connected to the local gas main, Spalding 1897. (Spalding Gentlemen's Society)

'How are things going?' asks Capt. Seymour Cooke, Charles Baldwin's event manager in Spalding, 1897. The balloon grows in size, attended by men from Spalding gas works. The man in the Stetson hat is the formidable landlord of the Black Swan Inn in whose field the event was staged. (Spalding Gentlemen's Society)

Above, opposite and overleaf: A sequence showing the vital job of unpacking the parachute, checking its suspension cords and shaking out the canopy to ensure there are no hitches prior to attachment to the balloon netting in Spalding 1897. (Spalding Gentlemen's Society)

Monsieur Auguste Gaudron, a dapper little Frenchman with a soft accent, piercing eyes and a neat waxed moustache, enjoyed a considerable international reputation as an aeronaut, claiming to have made over 700 ascents and descents in a career spanning twenty years – including being fished out of the sea (in 1894) in the best traditions of the old time balloon pilots!

Preceded by such a reputation and accompanied by his parachuting partner Miss Alma Beaumont, Gaudron was engaged to revitalise the Spalding show of 1 August 1898. Twenty-four-year-old Miss Beaumont – yet another of these Amazons – was a noted athlete, famous as a professional swimmer and high-diver, in addition to her parachuting prowess.

Auguste E. Gaudron, the eminent French aeronaut who made numerous flights from Spalding in the 1890s and married into the famous Spencer aeronautical family. (*Islington Gazette*)

Left: Fully inflated, Auguste Gaudron's balloon, *Alexandra Palace*, placidly awaits take-off time on the Black Swan field, Spalding, 1 August 1898. (Spalding Gentlemen's Society)

Below: Alexandra Palace lifts off at Spalding on 1 August 1898, clearly showing how Auguste Gaudron and Alma Beaumont and their parachute canopies are carried aloft. (Spalding Gentlemen's Society)

Gaudron's balloon was emblazoned with the legend *Alexandra Palace*, signifying his professional connection with that arena. In addition, there is both pictorial and written evidence of, what must surely be, the first appearance of aerial advertising in the region. Mr Matthew Herrod MPS, a Wisbech chemist and clearly a shrewd businessman, struck a deal with Monsieur Gaudron for the attachment of large banners to the balloon netting. These proclaimed the benefits of Mr Herrod's home-made pills and potions and in particular his (recently renamed) 'Balloon Lemonade Powder'.

Suitably adorned, the balloon carried Gaudron and Miss Beaumont aloft, taking twenty minutes to reach 8,500ft. Miss Beaumont was first to jump, and the ripple of apprehension buzzed around the arena below. Loud cheers, however, greeted the blossoming of her canopy and she drifted gently to earth but her leap caused a moment of drama for Monsieur Gaudron. The balloon lurched awkwardly at her sudden departure, and to the watchers seemed in imminent danger of overturning. No doubt this was due to the momentary loss of the pendulum-like balance. Cheers turned to gasps of alarm as Gaudron could be seen apparently struggling to jump, but in reality he was probably only trying to reposition himself on the seat to reduce the oscillation before jumping. The crowd's fear soon passed when Gaudron was seen to fall clear and land none the worse for his brush with danger.

The pair came down in Low Road, near the Chestnuts, home of Mr R. Culpin, with the balloon falling a little further away, about two miles from the arena. It is still surprising to record that, during the span of such events – nearly a century – not one person was injured in this region as a result of the random fall of a balloon! The final year of that Victorian century saw another close brush with death and once again it was the life of a young woman in jeopardy.

Auguste Gaudron, accompanied on this occasion by seventeen-year-old Miss Winifred Mansfield, returned to Spalding for the sports day of Monday 7 August 1899.

That erstwhile entrepreneur, Mr Herrod, took a large space in the *Spalding Guardian* newspaper to advertise his wares. This year all his products were rechristened 'Balloon etc', including such delights as Balloon Health Salts, Herb Extract, Lemon Crystals, Ginger Beer and Lemonade Powder. Such was his confidence in his Balloon Lemonade Powder that he distributed no less than 6,000 small, free sample tins on the day of the sports. Once again Mr Herrod negotiated advertising space on the balloon itself, with a canvas banner carrying 3ft tall lettering that would be visible to thousands all over the town.

Watched by a 10,000-strong crowd, massive for the size of the town, and swelled by special excursion trains, the balloon rose from the Black Swan cricket ground. Despite a smaller balloon being employed this year (14,500cu.ft), the two parachutists had no difficulty in gaining height. At a few hundred feet altitude, the drama began!

Miss Mansfield's parachute securing ribbon either broke or came undone prematurely. To the crowd's horror her canopy fell below the seat, billowing open to take the shape of an upturned umbrella. Was this to herald a sickening repetition of the Peterborough tragedy? The crowd held its breath.

It was impossible to escape as the balloon was now hundreds of feet in the air, but to their credit neither Gaudron nor Mansfield panicked. Frantically they hauled the offending canopy upwards against the pressure of the air. After a few heart-stopping moments they

One of the first examples of aerial advertising to be photographed. Auguste Gaudron's balloon in the Black Swan field, Spalding, prior to its flight on 7 August 1899, displaying an advert for Wisbech chemist Matthew Herrod's lemonade powder. (Spalding Gentlemen's Society)

Drama in the air! Winifred Mansfield's parachute seen billowing like an upturned umbrella beneath the balloon as it rises over Spalding town on 7 August 1899. (Spalding Gentlemen's Society)

succeeded in gathering it in; Gaudron turned the canopy right side up and, clutching the canopy to her bosom, the brave lady jumped, letting go of it as she fell. It worked!

Miss Mansfield landed safely after what seemed to be a normal gentle descent and was closely followed by Monsieur Gaudron, but as he remarked later – perhaps embellishing the story a little – instead of jumping at the usual height of about 6,000ft they had reached an estimated height of 12,000ft and both therefore drifted some distance apart. The lady came down on Cowbit Wash, at Handkerchief Hall, about two miles from the town, while Gaudron sailed on over the river Welland to alight on Deeping High Bank. Due to the height gained during the drama the balloon, on this occasion, they fell to earth rather more speedily than the parachutists at Locks Mill near Little London bridge.

Thus did Miss Winifred Mansfield escape with her life. She seems to have been none the worse for her brush with death, for this plucky lady – who also made a living as a variety artiste for twenty-eight years – turned up at Spalding a few years later with Monsieur Gaudron for yet another parachute descent.

The turn of the century heralded a reversal of fortunes when the August Bank Holiday sports day of 1900 was held on 6 August in appallingly bad weather. Auguste Gaudron was invited again and accompanied by Alma Beaumont this time; they were still being billed as the only double parachutists in the world. Public interest seemed to have heightened since that brush with danger the previous year but the intermittent, occasionally torrential, rain kept the crowd at the show down to a mere 3,600 stalwarts. High wind made it impossible to launch the balloon even though it had been inflated, and reluctantly the organisers cancelled that part of the event.

Over in Stamford that same day, Lord Exeter (in a scene to be repeated for B.C. Hucks and his aeroplane, twelve years later) was explaining in his pre-show luncheon address, that the Stamford Horse show was the first such event, combining and replacing earlier Bank Holiday racing sports shows in the town. Not to be outdone, the committee of the Stamford show and fête retained the services of Messrs C.G. Spencer & Son of London, to organise a balloon and parachute display for their 1900 inaugural event in Burghley Park, with the aeronaut and parachutist, Captain Alfred E. Smith.

Spencer's company had achieved considerable prominence in the field of balloon and parachute manufacture in the years around the turn of the century. This was due to the successful outcome of making a transition from participation in a rising cost activity, to manufacturing and selling equipment and professional services to those who could still afford the sport.

Just as at Spalding twenty miles away, the weather in Stamford on 6 August was equally dismal, putting the balloon's ascent in grave doubt. By 5 p.m., however, the bad weather cleared from the west, hastened by a rising wind. Captain Smith admitted that the wind presented risks but in true 'show must go on' spirit he decided to make the ascent.

With a hearty 'Goodbye all!' Captain Smith rose rapidly from the moorings, suspended beneath his 15,000cu.ft capacity balloon. It was carried quickly by the south-west wind, directly towards Belmesthorpe Meadows. Reaching 4,000ft in a now cloudless sky, Smith jumped from his seat, tumbling 300ft until his canopy fully deployed. Unable to affect his direction during his otherwise gentle descent, he was fortunate to miss, by a whisker, a ducking in the river Guash. With a final bump and somersault, Captain Smith reached terra firma in the parish of Borderville, a mere half a mile from his take off point. In less than half an hour he was being driven in a carriage through the park, greeted by hearty cheering and much handshaking.

No balloons were engaged for the 1901 August bank holiday celebrations in either Spalding or Stamford, but in 1902 the *Lincolnshire Free Press* once more carried huge advertisements for both the Spalding show and for that energetic Wisbech businessman, Mr Herrod. They contained evidence of the arrival of the twentieth century, by including a motorcycle race among its programme, in addition to Monsieur Gaudron's latest balloon show and Mr Herrod added 'Balloon Gingerette' to his growing list of drinks and tinctures.

As a further sign of the progress of aeronautics, and with an eye to expanding his commercial venture, Monsieur Gaudron now employed a 24,000cu.ft balloon with a wicker basket attached, capable of carrying four persons. Improvements had also been incorporated into the design of the balloon envelope by the addition of a sprung release valve at the apex, by which a gradual descent could be regulated by the 'pilot' remaining on board. Here, then, is evidence of a significant change in technique for these balloon events. In essence this stems from exploiting the passenger carrying potential, with parachuting being adapted to fit in with that new objective. It was, at least, an altogether more comfortable way for the parachutists to go about their business and was probably the closest one could get, with the prevailing technology, to the modern aeroplane drop. It was the intention of Auguste Gaudron to remain in the basket with his passenger colleague, Mr MacNeill, while their two parachutists, Miss Eva Hamilton and Mr Harry Grand, would jump from the basket at 5,000ft.

Inflation, supervised by Mr J.G. Hawkins, Spalding gas works manager, was completed by late afternoon and at 5 p.m. the majestic craft rose into a clear sky. Drifting in the direction of Spalding Marsh, it was not long before first one then another dot could be seen tumbling from the basket to be arrested by a blossoming white canopy.

With deft footwork, Miss Hamilton softened her contact with the roof of the Vine Inn, collecting only minor scratches before dropping into Albert Street at the rear of the pub. Harry Grand had a similar experience with a house roof a couple of hundred yards away in Willow Walk. There is no comment recorded about the feelings of the respective householders concerning the unannounced arrival of such hefty objects upon their roof, but one can guess!

With his new regulator to play with, Professor Gaudron descended very slowly near Weston railway station two miles from Spalding. Here, he and his colleague took the opportunity to carry out a few tethered flights for the benefit of the local people (and his own coffers of course) before packing up the balloon and returning to the sports field.

It was a monster 35,000cu.ft balloon, named *Vivienne*, that Auguste Gaudron brought to the town on 3 August 1903 and attendance at this, the twenty-seventh Spalding show, rose to 8,000 people. As in the previous year Monsieur Gaudron would pilot the balloon alone and his two lady parachutists, sisters Winnie and Ida Mansfield, would make the drop.

In a blustery wind, the large balloon climbed to a considerable height and when the ladies finally made their exit they landed at Moulton Austendyke, a good three miles or more from the arena. Even in this flat countryside the spectators would have had a pretty poor view of the descent but by now, perhaps, this may have seemed unimportant.

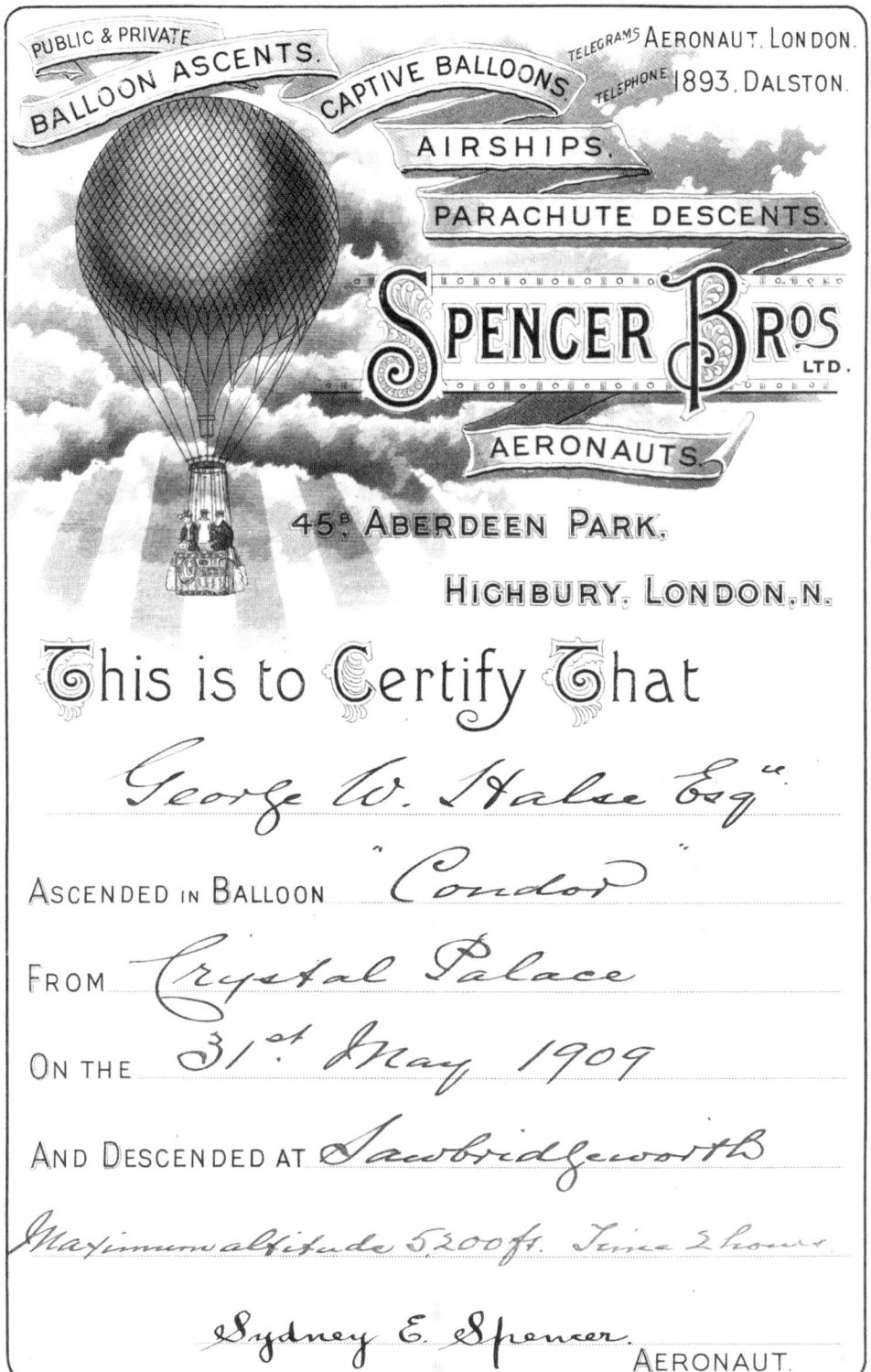

PUBLIC & PRIVATE
BALLOON ASCENTS.
CAPTIVE BALLOONS.
TELEGRAMS AERONAUT. LONDON.
TELEPHONE 1893, DALSTON.
AIRSHIPS,
PARACHUTE DESCENTS.

SPENCER BROS LTD.

AERONAUTS.

45ᴮ ABERDEEN PARK.

HICHBURY, LONDON, N.

This is to Certify That

George W. Halse Esqᵘ.

ASCENDED IN BALLOON *"Condor"*

FROM *Crystal Palace*

ON THE *31ˢᵗ May 1909*

AND DESCENDED AT *Sawbridgeworth*

Maximum altitude 5200ft. Time 2 hours

Sydney E. Spencer.
AERONAUT.

Certificate presented to a passenger in 1909 upon completion of a trip in a balloon operated by the famous aeronautics company, Spencer Bros Ltd of London.

By 1904, the C.G. Spencer Company of London had forged an enviable reputation in the field of ballooning and parachuting and as Auguste Gaudron was unable to attend this year the Spalding show was graced by the presence of Mr Henry Spencer. His responsibility was the staging of the exhibition and he acted as ground manager for the parachutists.

Bank holiday Monday 1 August 1904 in Spalding was blessed with brilliantly fine weather all through the morning and afternoon, encouraging over 8,000 spectators, to flock to the Black Swan field. The programme of events varied little from other years, including athletics, cycle races and even a tent-erecting competition by local volunteer soldiers. Immediately after the last athletics race, the grand event of the show took place. All day, in an adjacent paddock, Spencer's balloon, the 35,000cu.ft *Vivienne* from the previous year, had been filling steadily from an extension pipe linked to the gas main in New Road. This operation was under the watchful eye of Mr H.R. Wimhurst, manager of Spalding gas works. When buoyancy was achieved the balloon was manhandled into the main arena, being joined there by Mr Spencer and his two parachutists, Miss Winnie Mansfield and Mr Tommy Emms. Reverting to the trapeze-seat arrangement meant they would be unaccompanied and the balloon would deflate and fall to earth by itself after the pair had jumped.

Preceded by a sharp rain shower, enough to spoil the run of glorious sunshine, at 6 p.m. the balloon lifted from the arena. Drifting north towards Pinchbeck Fen, Miss Mansfield dropped safely near the town cemetery. Two minutes later Mr Emms leaped clear and landed without mishap at the Flax Mill, Pinchbeck. The balloon behaved itself and fluttered to earth, falling on Mr Burrell's farmland in Surfleet Marsh. An evening promenade with music in Ayscoughfee Gardens rounded off the splendid events of the day but another little shower put paid to the closing fireworks display.

'Professor' Gaudron excelled himself in 1905 by bringing his newest balloon, an immense 40,000cu.ft goliath, named *Zephyr*. In a boisterous wind, this monster required eight hours to fill with coal gas, this process being supervised by Mr Wimhurst the gas works manager, and Gaudron's colleague Captain Smith (Alfred E.?) at the usual venue behind the Black Swan pub.

With 'a handsome and comfortable-looking wicker car capable of carrying no less than seven persons', Monsieur Gaudron let it be known that he was willing to take up two or three fare-paying passengers. There is no indication that this offer was taken up or of the price asked for the privilege, therefore it cannot be deduced whether it was a high price or a fear of danger that dissuaded any takers.

It was announced that Miss Ida Cavanagh and Mr Tommy Emms would make the double descent. Miss Cavanagh, however, failed to arrive on time so Monsieur Gaudron insisted on taking her place in order not to deprive the show of its key element – the double descent. Captain Smith would remain on board in command of the balloon. At the very last minute though, Miss Cavanagh arrived in the arena amidst a burst of applause. She had, she explained breathlessly, simply missed her earthbound train! The descent was able to go on as originally planned and in bright sunshine the lady and gentleman floated gently to earth at Money Bridge, Pinchbeck West, without mishap.

With the size of balloons growing ever larger and flying to even greater heights, the dangers both in the air and on the ground were a constant source of concern to the promoters of the Spalding event, as indeed was the rising cost of staging the whole show.

The sports club committee, in its wisdom, therefore decided 'that some radical change is desirable in the programme, to give it a more attractive character'. Several suggestions were forthcoming with regard to introducing 'novelties'. For example, a quotation for an airship display had even been sought, although not pursued. Discussion abounded, much of it concerning finance, with the committee making it known that it could not carry on in the present way for much longer if it were to avoid the risk of financial loss. Furthermore, it was of the opinion that local tradesmen – who stood to benefit most from the crowds – should help to pay for the sports that attracted those crowds.

In respect of the balloon events, the committee declared, 'Although ballooning and parachuting has been instrumental in creating large attendances, it is regarded as dangerous and expensive and generally recognised as best omitted from future programmes.' Thus, in 1906, the main attraction at the August bank holiday show in Spalding was a gymnastic display and this was the pattern set for subsequent years, causing attendance to drop rapidly again.

At this stage in its history, ballooning in the showman-style had ceased to be attractive to audiences while it had also become unprofitable for its exponents. Ballooning was, like the monarchy at that time, now undergoing another change of direction. There were still occasional shows, fêtes and galas with balloons on their programme but these tailed off in the years leading up to the First World War.

Its heyday was, however, still far from over, for now ballooning was finding a new lease of life as a sport for a wealthy minority who had the money and leisure time to enjoy the freedom of the skies, unfettered in those days by complicated restrictions. Communications were effective and earthly travel by railway or ship was, for them, a comfortable and quite accessible means of reaching a desired start point or returning home afterwards. Coal gas, too, was now prolifically available in even the smallest towns.

This, then, was the new scenario for the sport. The old aeronauts moved with the times and became constructers, salesmen and tutors in the science of aerostation and the wealthy became their willing pupils. As an indication of the finance required to participate in the sport, there are reports of a ten-day holiday by balloon hired from C.G. Spencer & Son, costing upwards of £113 at this time. Clearly not a pastime for the working class!

Organisation was brought to the sport, too, with the foundation of the Aero Club of Great Britain in 1901 and, inevitably, it was not too long before thoughts turned to racing. Among such events, one emerged as the Grand Prix of balloon racing.

In 1906 an American, James Gordon Bennett Jr, playboy son of the owner of the *New York Herald*, gave his name to a trophy that the balloonists of the world could compete for on an international basis, with the winner to be decided by the longest distance flown from the launch site. Except for the interruption of wars and weather, the Coupe Aéronautique Gordon Bennett international balloon race has been a contest until the present day. Incidentally, this is the fellow whose outrageous lifestyle and involvement in newsworthy stunts – he was the one who sent Stanley to look for Livingstone – gave rise to that well-known expletive of incredulity, 'Gordon Bennett!' Chronologically, the next aeronautical event in the region was in fact the closing stage of this first Gordon Bennett race.

Sixteen coal gas-filled balloons started the race from Paris on Sunday 30 September 1906, competing for the honour of their countries. After a contest lasting twenty-eight

Balloons assembled in Paris for the start of the first Gordon Bennett air race of 1906. (Via Brian Cocks)

hours, victory went to Lieutenant Frank P. Lahm and co-pilot Major Henry B. Hershey of the USA in the balloon *United States*. Touching down near Robin Hood's Bay in Yorkshire, he was credited with flying a distance of 402 miles (647km). Under the rules of the race, the US Aero Club would hold the trophy for one year and must make the arrangements for the next race in America – if challenged. Lahm also picked up a £500 prize plus half the total entrance fees, representing another £80 for the winner.

Wouldn't Charles Green have rubbed his hands with glee at this type of contest?

Of those sixteen entrants, seven managed to cross the Channel and land in England, with the other nine descending on the continent. The second placed crew, Alfred Vonwiller and Ettore Cianetti, for Italy, flew 370 miles (593km) in *Elfe* before landing at New Holland, Lincolnshire, on the south bank of the river Humber, but it is the third placed balloon which is of particular interest to our story and will provide us with an insight into the conduct of balloon racing in those days.

Two of Britain's leading aeronautical exponents, the Hon. Charles S. Rolls (of Rolls-Royce) and Colonel J.E. Capper, entered with the balloon *Britannia*, a 78,500cu.ft giant made by Short Brothers. Lifting off from the Tuileries, Paris, at 4.20 p.m. that sunny Sunday afternoon, in view of the clear weather conditions prevailing, Rolls and his companion were – during an otherwise uneventful journey – able to chart their progress very accurately. Only after being becalmed over the river Seine for four hours, descending and rising in an effort to find usable air currents, were the aeronauts finally able to make headway towards the English Channel near Dieppe. At 11.20 a.m. on Monday, after more than nineteen hours in the air, *Britannia* crossed the English coast at Hastings. As a precaution, while crossing the Channel, they had 'trailed' a rope (Charles Green's invention) to maintain an altitude of 500ft, thus keeping in the northerly air current blowing at that height and to conserve ballast.

Drifting throughout the day, crossing Chatham, the Thames Estuary, Essex and Suffolk, by sunset on Monday Rolls and Capper passed over Thetford and Swaffham in Norfolk and came within sight of The Wash. Dropping small pilot balloons and watching smoke from chimneys, railway engines and fires, Rolls had detected a northerly air stream close to the ground. He hoped this would allow him to skirt to the west of The Wash and give him a good run to the north of England still with sufficient ballast and food for another day in the air if need be. Faced, though, with the inherent dangers to life and property on the ground by employing his trail rope over land to exploit that current, he reluctantly decided against this course of action. As it was, he was now in imminent prospect of drifting out over the North Sea and therefore decided to land. When Rolls made his decision, the balloon was over Shernborne, Norfolk and just beginning to pick up speed, bowling along at 35mph, a factor that settled the matter for the two travellers. They would have to land.

Britannia came easily to earth at 6.38 p.m. (in French, or race, time) in the flat, deserted coastal strip between Sandringham and The Wash shore, less than five minutes flying time from the sea and twenty-seven hours after take off. In a situation reminiscent of Charles Green's famous overnight flight from London to The Wash in 1835, there was much concern at the absence of communication with these two prominent gentlemen. As explanations clarified later, they were in excellent hands, being most hospitably entertained by the Reverend Waters, rector of Shernbourne, who put them up for the night. It had taken a good deal of time to deflate *Britannia* (only 10 per cent larger, though, than Charles Green's *Royal Coronation* in which he made his epic flight to Germany in 1836) and even longer to find a cart to carry it to the nearest village. By then, the hour was too late to send a telegram to the Aero Club in London, so they waited until Tuesday morning to do so. Meanwhile the rector, in addition to accommodating them, fulfilled a key task as the first suitable person to attest their competition documents – a strict requirement of the competition rules. Interviewed by the *Lynn Advertiser*, Charles Rolls said:

> The balloon was illuminated with electric light and [we] had an ample supply of hot beef and vegetables that were very comforting up in the dark clouds on a cold night. Of course no flame could be allowed near the balloon, so the food was heated on the 'Calorit' principle. Not long ago people did not care to go for a voyage in the air as they were sometimes much knocked about when the balloon touched earth. Nowadays, thanks to the ripping panel, the balloon was brought to earth gently. The ordinary valve is about a foot in diameter but the ripping panel, when opened, practically made a large slit from top to bottom, allowing gas to escape at once. It scarcely moved a couple of yards when we landed. The two cords for opening the two valves are different in colour and size so that by day or night there would be no mistake in handling them.

Early speculation on the result of the race suggested Rolls gained fourth place after Henri, Comte de la Vaulx and his companion Comte d'Oultremont for France in *Valhalla*, who also landed in Norfolk (at Walsingham). Rolls, however, was confident that the sealed

City of London, one of the UK entrants in the inaugural Gordon Bennett air race of 30 September 1906. Flown by Frank Hedges Butler and Percival Spencer, it was placed eleventh in the race. (Brian Cocks)

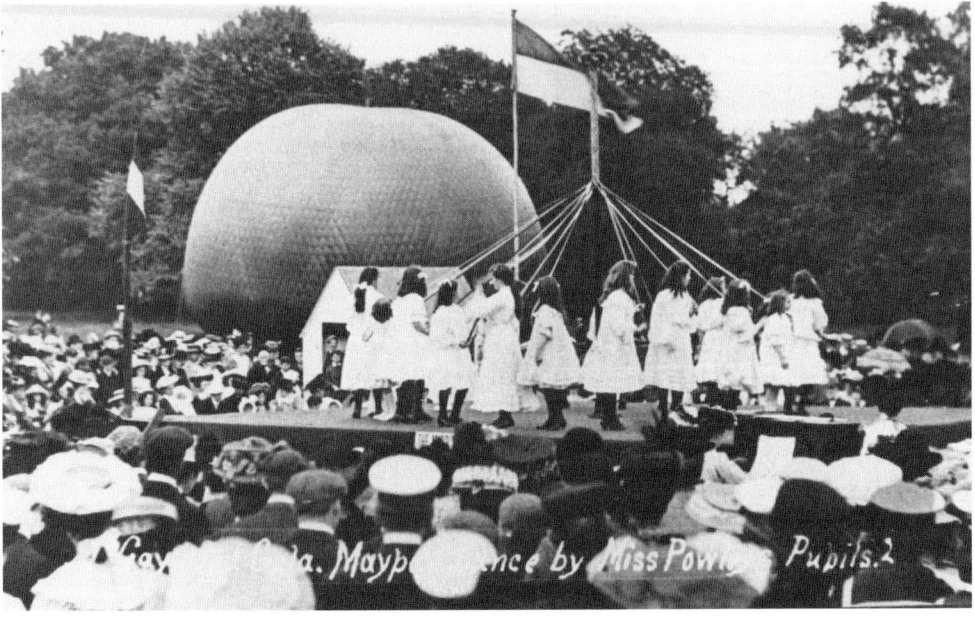

The balloon *Saladin* dwarfs Miss Powley's Maypole dancers at Gayton gala on 7 July 1907. Lt Lempriere flew the balloon from King's Lynn to Watlington.

instruments and his verified landing location would elevate him to third position. In the final analysis Rolls was proved correct and he was officially placed third by a whisker – *Britannia* logged 286 miles (461km) and *Valhalla* 285 miles (459km)! Britain's other entrant, *City Of London*, manned by Frank Hedges Butler and Percival Spencer, came down at Villers-sur-Mer, France, and was placed eleventh.

Despite the expansion of this aeronautical spectacle to an international scale, attention in the fenland region was refocussed at a more rural level for the King's Lynn gala of 1907. Held annually at Gaywood Park on the edge of the town, this event had all the ingredients of a traditional country fair, including 'maypole dancing by the young ladies of Miss Powley's private academy'. On Wednesday 7 July 1907, however, a contemporary photograph shows Miss Powley's young dancing ladies to be well overshadowed by the fat shape of a gas balloon lurking in the shelter of nearby trees.

Lieutenant Lempriere was booked to appear with his 30,000cu.ft coal gas balloon *Saladin* and to conduct a series of captive ascents with local people as passengers. High winds made the inflation process, begun on Tuesday evening, quite difficult – even in the lee of trees in the park. With a blustery wind persisting during the day of the gala and with seats, in the basket with the aeronaut, priced at 5 guineas (£5.25), no applications were forthcoming from anyone willing to undertake the thrilling trip. At 7.30 p.m., however, the wind abated and rather than disappoint the huge crowd, Lt Lempriere decided to make a solo free flight. With the bold lieutenant waving his hat energetically to the applauding crowd, the balloon veered off in the direction of Gaywood Hall and disappeared into the clouds. It was just half an hour after take off that *Saladin* came to earth safely at Thorpland, near Watlington, eight miles south of King's Lynn, where Lt Lempriere was able to partake of local hospitality when he was invited to spend the night at Thorpland Hall, from whence he returned to Gaywood next day.

By 1907 then, the swansong of ballooning, at least as a public entertainment for shows, fêtes and galas in the fenlands, had arrived and the last act was played out in Spalding a month after the Gaywood gala. Bank holiday Monday 5 August 1907 was Auguste Gaudron's farewell appearance at the Spalding show, and no public celebration was really complete without a show of fireworks and a rousing musical evening, and that is as true today as it was a century and more ago.

For the culmination of Spalding's Bank Holiday show, in 1907 Auguste Gaudron lent his considerable expertise to the evening programme rather than a balloon exhibition during the day. He was invited to be the star attraction of the evening's entertainment in Ayscoughfee Hall gardens where his balloon was inflated in an adjacent paddock.

At 9 p.m. an interval in the music was announced, whereupon 'the event of the evening' took place. This year the show committee, lead once again by Tom A. White, introduced a novel approach to the traditional firework display. This took the form of an ascent by the celebrated Professor Gaudron who, accompanied by Mr A.E. Hallas, editor of the *Spalding Guardian*, went aloft with a light framework hanging beneath the car of his balloon to which were attached fireworks, to be set alight at a suitable height above the town for all to see.

Gaudron must have felt pretty confident about his equipment, for the combination of a balloon filled with coal gas and fireworks going off in close proximity was a recipe for disaster if ever there was one! In fact it was an injudicious mixing of fire and gas scenario that caused the death of the famous Pilatre de Rozier back in 1785. In the event there was no disaster on this occasion and the fireworks display was visible for many miles across the fens, as far away even as Bourne, twelve miles distant.

Without elaborating on detail, the *Spalding Guardian* observed, 'owing to an accident, this feature was not the success anticipated'. Perhaps Monsieur Gaudron and Mr Hallas were lucky fellows indeed in being able to complete the display, but they escaped with their lives and came back to earth safely on the Spalding to Weston Road, two miles east of the take-off point.

Whatever the little 'accident' was, it seems to have been the final straw for the Spalding show committee. It was too close a brush with disaster to run any more risks. Gaudron and Hallas, therefore, had made the last coal gas-filled free balloon ascent to be seen in Spalding.

It fell to the French, however, to bring a century of free ballooning in the region to a close. We have already read of the influence of Gorden Bennett, and balloon racing rapidly grew in popularity. Many newspapers in the region reported sightings of brightly coloured balloons passing over eastern England on Monday 16 June 1913. These were later discovered to belong to participants in a long distance race, organised by L'Aéro Club de France that began in Paris St Cloud the previous afternoon. Scotland was the stated objective of the competitors but even though the wind in France held fair, none of the starters got further than Lincolnshire. The race winner was Monsieur Pierron in balloon No.9 that landed at Fotherby, near Louth. Having lifted off from Paris at 3 p.m. the previous day, Sunday 15 June, he and his companion made an uneventful crossing of the Channel, then their course took the balloon via Newhaven and Brighton north to Bedford, Peterborough, Spalding and Boston, reaching a maximum altitude of 20,000ft during its journey. Monsieur Pierron decided to land when it became clear that the favourable wind had died away completely and he was making no further northerly progress whatsoever. His balloon was deflated and carried on a dray into Louth from whence he and all his accoutrements set off by train for the journey back to Paris.

Several of the twenty-two contestants came down in Lincolnshire and adjacent counties. In the crystal clear sky, one of these, No.16, was spotted passing high over Downham Market and Ely, heading north-west. This was piloted by Monsieur Maurice Bourgeois, current holder of the Club's Grand Prix, with his companion, a South American lady, Mme M.L. De Pulido, owner of the balloon. They landed at Skellingthorpe, Lincoln, and a reporter from the *Lincolnshire Echo* was soon at the scene:

Despite the invention of airships, the balloon still manages to fascinate onlookers when it is in flight. It approached Lincoln at a great height but the air was so clear that French flags could be seen fluttering on its side. It descended gradually and came to rest on Mr Green's farm in Skellingthorpe where a large crowd quickly gathered. Just before the landing, in response to a message shouted through a megaphone by the lady passenger, members of the crowd caught hold of trailing ropes and held it steady until the occupants were able to alight. It was then anchored, packed and conveyed to Lincoln. Later the lady and her companion were taken by motor car to the Great Northern Hotel in the city. Before leaving Skellingthorpe, local residents Mr and Mrs E H Neville offered every assistance and Mr Neville being a JP signed their competition certificate to confirm the landing site, which gave them second place. Madam Pulido, who spoke English perfectly, said she and her pilot crossed the Channel at 11.30am and had travelled straight up the country heading for Scotland. When they reached Lincoln, however, it was found impossible to travel any further north, there being scarcely a breeze at the time.

It turned out to be quite a busy aviation weekend for the people of Louth when another balloon was seen passing over the town, 'causing much excitement despite the fact that they had witnessed aeroplane demonstrations on Friday and Saturday by the well known aviator Mr B C Hucks'.

This one was piloted by Mons Alfred LeBlanc and his wife, placed third when it was brought to earth at Marsh Chapel near North Coates in order to avoid the prospect of drifting out to sea. Not all competitors were as fortunate as the LeBlancs though. The crew of a Dutch tugboat rescued two Frenchmen clinging to the wreckage of their balloon in the sea twenty-two miles south of Ventnor, Isle of Wight. Meanwhile Messrs Dubois and Spire in a balloon marked 'Aero Club De Touraine' landed at Ayston Hall near Uppingham, even as another balloon arrived in Lincolnshire:

> The balloon was travelling from the direction of Stamford and had descended to tree-top height, trailing a rope. Two small boys caught hold of the rope but were lifted off their feet and into a hedge before they let it go. Shortly afterwards it came down in a grass field without injury to the occupants. The leader of the party, a Frenchman, was taken to the village and entertained to lunch. He borrowed a cycle and rode to Thurlby railway station where he made arrangements to have his balloon collected and transported home. A villager provided a horse and wagon to transport the balloon to the station.

With it went the era of free-flight gas ballooning in the eastern region – at least for another seventy years, until hot-air ballooning had its renaissance – and then – Gordon Bennett! That man's balloon race came over again! 2006 was the centenary year for the Coupe Gordon Bennett. During the intervening one hundred years, race venues moved around the world according to the nationality of the winners, and as such there was no venue that brought the competitors within sight of the UK – until now.

The seventeen teams taking part in the 2006 anniversary race – the fiftieth actual race – set off from Waasmunster in Belgium on 9 September. David Hempleman-Adams and Jonathan Mason were in a German-registered hydrogen-filled balloon, D-OWNT, representing Great Britain, gaining third place and one of several balloons tracking over eastern England. Winners of the event were Belgians Phillipe De Cock and Ronny Van Havere in D-OCOX, *Belgica 2*, one of three balloons entered by that country. Airborne for sixty-seven hours they covered 1,522 miles (2,450km) at a speed over the ground of 32mph (52kph), landing near Kirkenes in northern Norway at midday on the 12th.

By 1912 the attention of the public here in the region, as indeed in the country at large, was being captured by a new phenomenon – the aeroplane. With the arrival of the internal combustion engine the free balloon, after more than 100 years, seemed to

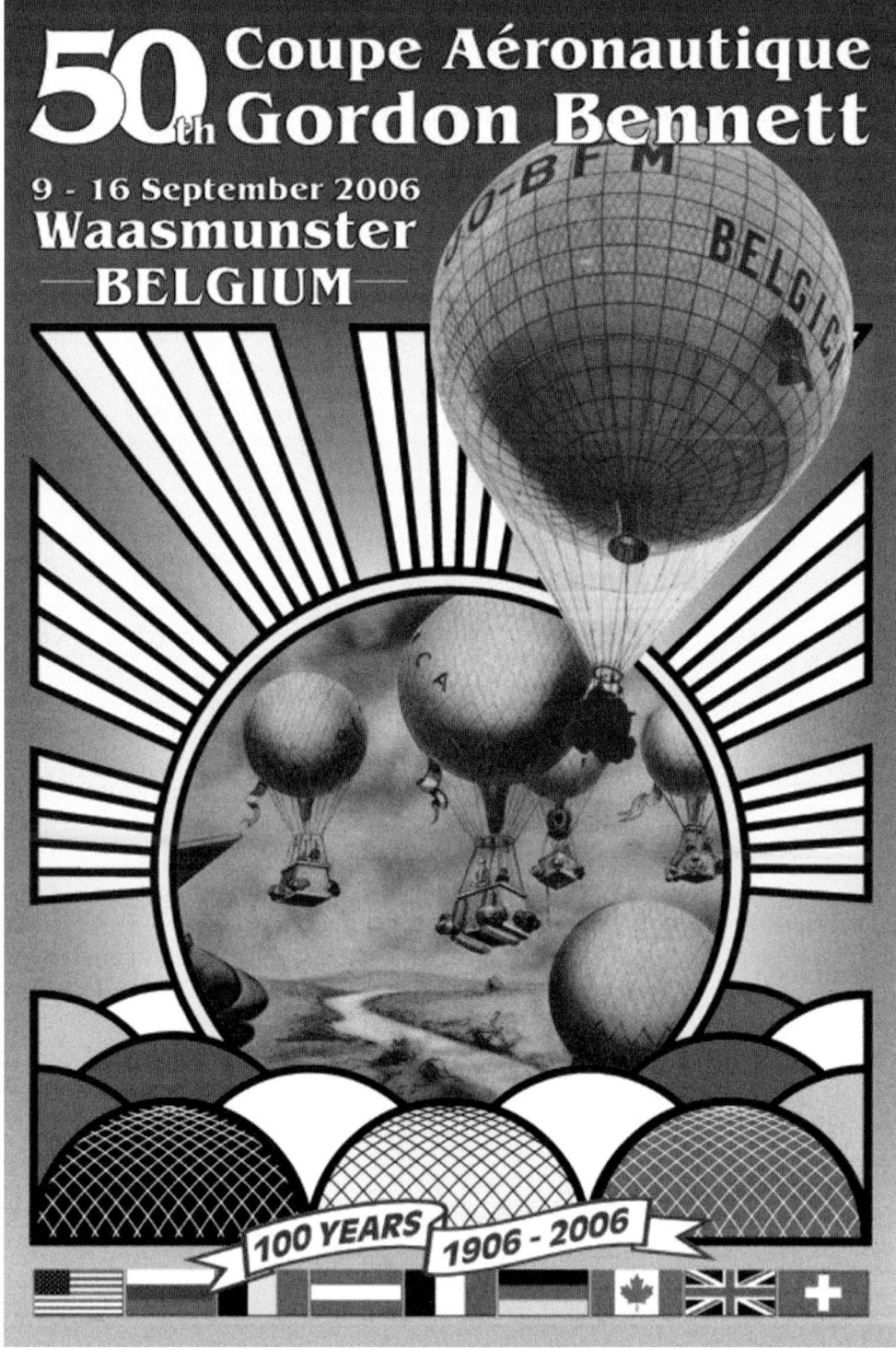

Official logo for the 2006 centenary Gordon Bennett balloon race. Starting in Belgium some contestants overflew eastern England during the race. (The Coupe Gordon Bennett organisation)

have become an anachronism. In more than one sense, ballooning was going nowhere! It would remain so for another fifty years, until first scientists then sportsmen gave it new life in the realms of space research, meteorology, astronomy and – once again – pleasure.

But now to begin the aeroplane's stirring contribution to this region's rich aeronautical history.

CHAPTER THREE

Those Magnificent Men...

To William Hugh Ewen goes the distinction of being the first person to bring powered flight to a fenland audience.

This momentous event took place on the dull and windy evening of Friday 28 June 1912 near the little village of Walton on the outskirts of Peterborough, when an immense crowd, of between 15,000 and 20,000 people, flocked to the landing ground in Marholm Road. There, spilling over into the surrounding meadows and by-ways, they waited expectantly for a glimpse of this new wonder of the age.

Following the publicity of Louis Blériot's cross-Channel flight of 1909, there was a surge of interest in learning to fly aeroplanes. It was not long before schools for this purpose sprang up all over the country, run by men who themselves could really only claim a rudimentary knowledge of flying.

Born in Shanghai on 1 December 1879, of Scottish parents, William Hugh Ewen had a rich academic early life in Scotland. After his schooling at Heriot's College, Edinburgh, he went to Edinburgh University to study medicine. Deciding that a medical career was not for him, he left to undertake training in business. This kept him occupied for the next five years, after which he returned to the university to study music, a passion for which he had held since his early childhood. Having gained a first class degree in music and a reputation as a brilliant organist, he held a number of posts culminating in appointment as organist of Queen's Park church in Glasgow.

It was during this latter period, covering four years from 1906, that he became seriously interested in the science of aeronautics and carried out private experiments with model flying machines. William found his niche in life, though, when he moved up to the real thing.

Ewen was awarded his Royal Aeronautical Club (RAeC) flying certificate, No.63, in 1910 after a mere three days of tuition from the famous Louis Blériot himself, at his Hendon flying school – a record at the time. While gaining more flying experience during that year, he recognised the business potential of what was still regarded as a sport for the wealthier classes. So, returning from London to the family home in Lanark in 1911, he set up a flying school (no special credentials were necessary) on the racecourse outside the town. Equipment for his school, the first in Scotland, was initially one Blériot monoplane but he soon augmented this with a biplane. Both aeroplanes were assembled in a hangar under his supervision and he expressed his intention to construct an aeroplane of his own design before too long.

Although flying was all the rage, the financial success of such schools was by no means a foregone conclusion. For every new school opening, there were others feeling the pinch of

William Hugh Ewen, pioneer aviator and musician. He was engaged by the *Daily Mail* to tour eastern England in 1912. (Via John Knight)

the rapidly escalating costs associated with this new dimension of travel and closures became equally common. It therefore took an enterprising pilot with panache, business acumen and not a little courage, to earn a living from flying in those pioneering days. William Ewen was one such man and his school survived. However, in order to further exploit the burgeoning flying training market, Ewen recognised that he would have to venture south of the border and so he transferred his school to Hendon Aerodrome in north London.

One of the most effective ways of gaining both flying experience and enhancing one's personal reputation was to participate in the rapidly expanding calendar of flying meetings and air races. William Ewen was a frequent participant in the Hendon meetings and other similar events, from April 1912 onwards. Soon he acquired a national reputation, according to *Flight* magazine, as 'a pilot of great courage and ability'. Such qualities would come to stand him in good stead on many occasions in the air. His Hendon school flourished as a result and he was able to buy a variety of aeroplane types, including a Deperdussin, a Blériot and a Nieuport, on which his aspiring pupils might learn to fly for the princely sum of £75, including, according to his advertisement, breakages!

Shortly after the inaugural Hendon meeting, Ewen bought a Caudron biplane. He was so impressed by the quality and robustness of this nippy, 25ft wingspan machine, that he became the sole distributor in the UK for this type of aeroplane.

It was early in 1912 that the *Daily Mail* newspaper, ever keen to stimulate the nation's interest in this new technology, while promoting its own sales of course, sponsored a 'Grand Aviation Tour of England'. The newspaper engaged Monsieur Henri Salmet, regarded by many as the leading pilot of the day, to tour parts of rural England with his aeroplane, 'to give countryfolk an opportunity to see first-rate demonstrations of flying'. Such was the success of Salmet's tour of the West Country that the *Daily Mail* engaged more pilots to cover other parts of England. The *Daily Mail* reported:

> The demand in the North to see the *Daily Mail* airmen has been so great that it has been found impossible for Mons Salmet to make the complete tour of England that was proposed. 'The Flying Scot' Mr W H Ewen will … make an aeroplane tour of the northern counties. He starts north from Hendon tomorrow, weather permitting, for Cambridge. The people of Peterborough will have the opportunity of seeing the airman on Thursday. If it is good flying weather Mr Ewen hopes to go on to Lincoln on Thursday afternoon. Hull may expect to see him on Friday.

William Ewen was the first of this new team, and in late June 1912 left Hendon to tour the East Midlands and north of England where his role as a 'Daily Mail Airman' now brings us back to that expectant throng in Walton.

An air of bitter disappointment, left over from the previous day, pervaded the district. One London newspaper had misled the local populace into believing the aviator would arrive on Thursday. In the words of the *Peterborough Advertiser* the next day, 'they swarmed [to Walton] a-foot and a-wheel, in motors, taxis, cabs, landaus, traps and the electric cars reaped a rich harvest'.

Entrance to the proposed landing field, near the Walton and Marholm railway crossing, was sixpence (2½p) a head and over 2,000 paid to enter that afternoon and evening. They were encouraged to part with their money by the tour advance manager's claim that 80 per cent of the takings would go to expenses and 20 per cent as a donation to Peterborough Infirmary.

As 9 p.m. approached it was clear that Mr Ewen would not show up that day and the massive crowd shuffled off home, thankfully for the hard-pressed small police contingent, in a generally jocular mood. Those who had paid to enter the field got their money back so a potentially awkward situation was avoided. Ewen had flown from Hendon to Cambridge and planned to proceed to Peterborough on the morning of Thursday 27 June, but a strong wind pushed him off course to the west and he was forced to land at St Neots in pouring rain. It was not until 8.30 p.m. on Friday 28 June that the wind dropped sufficiently to allow him to complete his flight to Peterborough.

Meanwhile, word spread like wildfire that he had taken off, and the city emptied once more; at 8.50 p.m. a cheer went up swelling in volume as first one then others glimpsed a speck in the evening sky. Slowly that speck took the shape of an aeroplane and the clatter of its engine could be heard above the buzz of the crowd. Mr Ewen flew in a gentle sweep around the landing ground, across the railway line and over Werrington village before gliding gently to earth in the meadow, all to the obvious delight of the crowd.

The show begins. William Ewen is about to take off in his Caudron. Peterborough 1912. (*Peterborough Advertiser*)

The flight from St Neots to Walton had taken twenty-eight minutes at a height of 2,000ft. Ewen was the man of the hour and carried shoulder high to a waiting motor car to be whisked away in regal splendour to a Peterborough hotel for the night.

Saturday 29 June dawned bright and clear and William Ewen gave two flying exhibitions during the day, motoring later to Lincoln to select his next landing ground. One of the pleasures of being a celebrity was that he was fêted at every venue, and to mark this auspicious occasion Councillor C.E. Crawley, a former mayor of Peterborough, presented Mr Ewen with a splendid silver rose bowl on completion of the flying display. In his short speech of welcome the councillor remarked:

You are, I think, the first gentleman to land here from the sky [loud applause]. I remember the old coaching days and the entry of the first railway into Peterborough and I remember how wonderful we thought of them to have coaches moving without horses. Now you flying engineers [sic] can come even through the air. I wish to thank you for the great pleasure your clever flying has given us and ask you to accept this rose bowl as a memento of this happy occasion.

Responding, Mr Ewen said:

Aviation is a seriously useful and practical science. I should be very sorry if you looked upon it only as a show. It is more than that. Its progress means more to our country than that. Of course, flying is a costly business and no aviator could undertake demonstrations on the wide lines that I and other aviators are now making, without some very ample support, such as is now being given me by the *Daily Mail* [more applause].

PRESENTATION TO DAILY MAIL AIRMAN AT PETERBOROUGH

Above and below: Presentation of a rose bowl to William Ewen by the mayor in front of the Caudron biplane and an expectant crowd in Marholm, Peterborough, on 29 June 1912.
(*Peterborough Advertiser*)

DAILY MAIL AIRMAN AT PETERBOROUGH

It was at 5.30 a.m. the following Monday 1 July, when he started his engine in preparation for the flight to Lincoln but it did not run smoothly, as one of the cylinders of the 60hp Anzani motor was not sparking properly. On this occasion Ewen's mechanic was to fly with him and since this Caudron was only a single-seater, it meant the poor fellow had to lie on top of the fuel tank in a somewhat precarious position – but it was something he had done before.

With the engine still coughing Ewen rashly decided to take off but almost inevitably, with the misfiring and the extra weight of the mechanic, he had difficulty in gaining height. Faced with the prospect of a collision with the trees of Pocock Wood, he turned quickly and seems to have got into a classic stall situation from which he crashed. Fortunately the aeroplane was neither high nor travelling at any great speed so both Ewen and his mechanic were able to scramble clear, bruised and shaken. There is no record of any comment made by the mechanic!

The aeroplane suffered a broken wing and propeller, and two of the six engine cylinders were broken off. Undeterred, Ewen had the aeroplane removed to a barn in Gunthorpe where the two aviators carried out repairs that were completed by the following weekend.

This time there was no hitch and William Ewen set course – alone since Mr Warren the mechanic, no doubt chastened by his previous experience, opted to go by road – for his next venue, Lincoln, at 6 p.m. on the evening of Monday 8 July 1912, landing at Wragby Road at 9 p.m. after yet another strenuous flight.

Following the A15 road northwards he flew over Bourne at 6.30 p.m. but then a crosswind got up and clouds began to roll in across his track making it difficult to maintain his course. Chugging along at 500ft, Ewen struggled on past Sleaford and Ruskington at 6.45 p.m. when finally his good sense prevailed and he landed between Scopwick and Digby – much to the delight of the villagers, who urged him to stay the night – to check where he was and to see if the visibility might clear sufficiently for him to get to Lincoln that evening. He was only too aware that everyone in Lincoln was getting rather impatient with his delays, even though much of it was weather related. After a couple of hours the sky cleared but he realised if he was going to make Lincoln with the last of the daylight he had to go now. So, with the cheers of a couple of thousand people – the first aeroplane in the area had that much drawing power – fading behind him off he went again.

Lincoln's first glimpse of the Caudron was as it came in from Branston over South Common, sweeping round the west of the city towards the grammar school field. In the gathering dusk Ewen missed the field and it was only when rockets were fired that he spotted it and came in to land. Continuing his somewhat erratic progress he made a circuit of the field and came in, throttle closed, for his landing. In the fading light Ewen made a hash of it and was catapulted out of the cockpit as the aeroplane undershot into a barley field, hit a fence and partially overturned. Although dazed, he was helped to his feet by a doctor and a policeman then, like the trouper he was, his first thought was to inspect his machine. The propeller and tailplane were damaged but satisfied all was repairable, he hobbled over to the welcoming party amidst lusty cheers from the many thousands who had turned out to see this 'sight of a lifetime'.

Once again Ewen had suffered bruising and a sprained ankle but no doubt felt it worthwhile when he was presented with yet another dazzling gift to mark the occasion. The mayor of Lincoln made a speech of welcome and presented the intrepid aviator with a gold scarf ring in the shape of a Lincoln Imp studded with diamonds, the result of a collection by prominent businessmen to remind him of his visit. This was perhaps quite

Above and below: Oops! William Ewen's broken Caudron after its mishap due to engine failure in Peterborough, 1912. (*Peterborough Advertiser*)

useful since by now he could probably not remember which of his bruises had been sustained at which venue! It is interesting to note that Ewen carried a small bag of post on his flight from Peterborough to Lincoln. Postcards celebrating the event were sold in newsagents in Peterborough and buyers simply wrote an address on the card and handed it back to the newsagent.

Retiring to the Saracen's Head Hotel for the night, it emerged the following morning during numerous interviews that the repairs needed made it unlikely a flying display could be put on for at least two days. He also changed the venue for the display to West Common because the field in Wragby Road was, in his opinion, too small.

Bad luck continued to dog William Ewen's aerial progress. With hundreds of people already assembled and thousands more streaming towards the Common, he and Mr Warren, his mechanic, made a final inspection of the Caudron. But the crowd was not the only thing gathering; ominous black clouds drifted over the city and just as the pre-flight checks were completed there was a series of thunderclaps and the Heavens opened with an almighty deluge. Lightning lit up the bedraggled scene and the crowd scattered in search of shelter from the storm.

In danger of being wrecked by either the howling gale, torrential rain or lightning – or all three – it was only due to the brave efforts of Mr Warren and a little group of helpers, all soaked to the skin, who clung on to the aeroplane, holding it down in the teeth of the wind. With no time to cover the machine with its usual large tarpaulin Warren borrowed a policeman's cape and threw it over the engine to keep the water out. The storm passed but not without leaving the Caudron's rigging and struts damaged or strained. Flying was now quite out of the question.

Hugh was disconsolate over his bad luck but more disappointed because he could not fulfil his promise to entertain the crowd with a flying display before setting off for Hull that evening. At crack of dawn next morning the mechanic and pilot were up and making repairs. These were complete by 7 a.m.; Ewen made a few ground runs along the common to test everything and he declared the aeroplane ready to go. It was still not to be. A stiff wind sprang up and Ewen decided it was now far too strong to risk the flight. Once again the crowd dispersed, disappointed, to await his next attempt. They would wait in vain. With no sign of the bad weather abating Hugh finally gave up any chance of flying to Hull and, anxious to keep to his commitment in that city, he left Lincoln by car, leaving arrangements for the Caudron to be dismantled and taken to Hull by lorry.

William Ewen carried on with his flying business right up to the start of the First World War. He then served his country with distinction, helping to organise flying training in the Royal Flying Corps and later the Royal Air Force, rising to the rank of major (squadron leader). After the war he returned to his first love, that of music and became musical director of many successful stage shows throughout the inter-war years. He died on 26 November 1947, and what greater compliment can be paid to him than to say that in his lifetime he had played a valuable part in the development of aviation in Britain.

The crowd-pulling potential of aeroplanes and their pilots – often regarded as 'supermen' – was uppermost in the minds of the organisers of the 1912 Lincolnshire Agricultural Show. For the first time, the show venue was in Skegness and its committee was anxious to create the right atmosphere for success. For this reason special attraction plans for that year included Mr John Brereton and his Blackburn monoplane. His brief was to put on a flying exhibition each day and to make himself available 'to explain the workings of the machine to the crowd'. The show was held at Skegness from Wednesday to Friday, 17–19 July 1912, and after a slow start a total of 18,000 people passed through the turnstiles over the three day period.

John Brereton and his aeroplane actually arrived by train from Filey, Yorkshire, once again indicating that, in those days, rail travel compared to air travel was more reliable and not subject to the vagaries of the weather. Filey was the home of Robert Blackburn's aeroplane construction company at that time.

Like many of his contemporaries, John Brereton was a pioneer motorist as well as an aviator. Born in Ireland in 1884, he travelled extensively in his youth before settling in London in 1900 to work as a motor engineer. By 1911 Brereton had developed a keen interest in aviation, learning to fly at Brooklands and receiving RAeC certificate No.136 in September of that year. His flying course cost him a total of £175. In October 1911 he obtained a post as pilot with the Fritz Gortze Monoplane Company in London but the heavy burden of expenses associated with flying, even in those days, ruined the company and Brereton left to join Blackburn at Filey.

During his time as pilot/manager with Blackburn he toured the country giving exhibitions of the company's product and speaking publicly on the science of flying and this accounts for his presence at Skegness. Sadly the dull, wet weather over the period of the Lincolnshire show allowed only one short flight on the second day. He took off in light rain and the Blackburn monoplane described a graceful circle out to sea, returning

John Brereton about to take off in Blackburn Mercury III from Bridlington beach on 15 July 1912. He brought this Blackburn aircraft to Skegness the next day. (G. Stuart Leslie)

to earth just five minutes later, without mishap. The expectant crowd was disappointed but conditions were declared too dangerous for further flights.

John Brereton left the employ of Blackburn later in 1912 for the British Deperdussin Aircraft Company, joining it as a manager/instructor. However, like Gortze, Deperdussin too succumbed to mounting costs and was wound up in February 1913. Brereton remained in flying and sailed to Canada to try his luck there but finding the state of aviation inferior to that in Britain, he entered the USA and found a job exhibition flying in Los Angeles.

James Brereton, as he appears in his pilot licence photograph. (RAFM)

Bentfield Charles Hucks, the epitome of
pre-First World War exhibition aviators.
(G. Stuart Leslie)

Upon the outbreak of the First World War he returned to England, joining the RNAS as a flying instructor at Eastchurch. Discharged as medically unfit in 1915, he recuperated then became a civilian test pilot at the Royal Aircraft Factory, Farnborough. Brereton rejoined the armed forces in February 1916, this time as an NCO, instructing air mechanics in the RFC. Eventually he regained his pilot certificate that month, finishing his war service with a commission and the post of chief engine specialist at RAF Henlow.

During his flying career John Brereton was chief flying instructor at RAF Waddington, and it is through his association with that station that he decided to settle in Lincoln after the war, taking up his trade as a motor engineer once more. Brereton died in Lincoln in November 1936 at the early age of fifty-two, another who had done his bit for aviation.

The names of Ewen and Brereton were in the public limelight for but a brief period until, even by the beginning of the First World War, aeroplanes had become such an accepted part of life that their names were lost among the 864 pilots who had qualified prior to August 1914. Their contribution to British aviation, though, should neither be underestimated nor forgotten.

However, from among their contemporaries emerged a pilot whose reputation grew steadily, so that by the end of an all too short life his became a name much revered in aviation circles.

Bentfield Charles Hucks, known to the public as B.C. Hucks and 'Benny' to his friends, became a regular visitor to East Midland skies in those far off days before the First World War.

The first occasion arose as the result of a personal invitation from Lord Burghley himself, once again underlining the high regard in which these early aviators were held by all levels of society. Reflecting the fashionable interest in aeronautics displayed by his ancestor a century earlier, the Marquess of Exeter, patron of the annual Stamford Horse and Poultry Show, held in the grounds of Burghley House, gave his support to an air show to promote the event. It was his custom to preside over a pre-show public luncheon attended by helpers, local dignitaries, business and prominent people. In the year 1912 the show, described as part of a 'great Unionist [political] demonstration' was held on 12 September and Mr Hucks was in attendance with his Blériot aeroplane.

In his speech Lord Burghley said ' … I … believe it is the first time the show has included among its attractions poultry, cows and a flying man … [laughter]. In respect of livestock classes, as far as flying men are concerned … I do not suppose we would be able to arrange a separate class for them [more laughter]. There is only one entry and I hope Mr Hucks will have a very pleasant flight over the area of Stamford [applause].'

In fine summer weather Hucks did not disappoint and put on a spirited flying exhibition during that afternoon and helped draw a huge crowd of visitors to the event.

In little over a year Benny Hucks, now twenty-eight years of age, had become a star attraction at such venues all over Great Britain and was generally credited as being the first pilot to loop-the-loop and to fly upside down under control – the latter being the highly relevant bit in those days!

Born in Essex in 1884, the son of an engineer, Hucks followed his father's footsteps into the engineering profession but after completing his apprenticeship he was bitten by the aviation 'bug'. Moving to Blackpool in 1910 to pursue his new career, he joined the Claude Graham-White Co., servicing aeroplanes for one of the founding fathers of British aviation. Benny soon acquired a taste for flying and was much encouraged by 'G-W' who gave him flying lessons. Hucks was quick to master this new dimension.

Towards the close of 1910 he decided to move on once more and became another of those aviators to recognise the potential and rapidly growing reputation of Robert Blackburn, joining him at his workshop in Filey, on the Yorkshire coast.

Blackburn had recently built his first flying machine, a monoplane christened *Mercury*, but it was beset with engine problems. Blackburn and Hucks shared the air-testing between them using Filey Beach, near Hunmanby Gap, as their flying ground while trying to improve the engine's performance. By early 1911 Benny's skill as a pilot began to emerge and eventually he took over all the flight testing for *Mercury*. This, incidentally, was before he had actually gained his official RAeC certificate!

This certificate, sometimes called the Brevet, was part of a system of standards introduced in Britain by the Royal Aeronautical Club (RAeC) in March 1910. From that date the RAeC undertook responsibility for setting standards for the assessment of flying competency. They issued certificates to civilian and military candidates having passed a series of flying tests. The club modified these tests as time passed and experience grew. Originally, three separate flights of three miles each were required. These were to be

Above and below: B.C. Hucks at Burghley Park, Stamford, 12 September 1912. (Via David Robinson)

round a circular course, without an intermediate landing, and each final landing was to be made with the engine stopped. This latter approach procedure was called volplaning and later became known as a 'dead-stick' landing.

It was on 30 May 1911 that Hucks eventually took his own certificate tests on the Blackburn Monoplane *Mercury* at Filey. The programme had been modified that year so that he was now required to complete two 5km flights in a closed circuit. These flights were to consist of not less than five figures-of-eight around two poles set 500m (550 yards) apart. This was to ensure that the budding pilot could control the aeroplane under some pressure in both directions. The second part of the test required Hucks to make a separate flight during which he was to climb to a height of no less than 500 metres (1,650ft). At the end of all these flights the candidate had to land with the engine switched off, having to stop within 50m (55 yards) of a point on the ground previously nominated.

Benny Hucks successfully completed these tests and was awarded RAeC certificate No.91. His test, however, was not without incident, for as he landed after one of the prescribed flights the propeller of the Blackburn Mercury fell off!

Parting company with Robert Blackburn in 1911 he became self-employed as an exhibition pilot, earning a living from racing prize money, sponsorships, product endorsements (he featured, for example, in Phospherine Tonic newspaper adverts) and exhibition fees. He was in his element and the darling of the public. Thus when the *Daily Mail* sought to increase the number of pilots on its publicity tour, Benny Hucks was a natural choice to join the team.

Being an engineer, Hucks designed many gadgets for his aeroplane. During his 1912 tours to the East Midlands, East Anglia and Lancashire he demonstrated an electrical release device for bomb dropping, using thunder-flash bombs. His airborne 'searchlights', one at nose and tail, were a regular feature of twilight flights and he was engaged in a novel way at the 1912 Midlothian (Scotland) election when he distributed leaflets by air during the campaign. When the *Daily Mail* continued its publicity tour into 1913, Hucks once again visited the East Midlands, this time taking in Peterborough, Stamford, Spalding, Boston, Skegness, Sleaford and Lincoln.

Earlier visits to the region had been essentially solo exhibitions. With the arrival of Mr Hucks onto the aviation scene, the public became more actively involved. He offered limited opportunities for passenger flights, in addition to bomb dropping demonstrations, leaflet dropping, aerial fireworks and searchlight displays. Clearly these innovations enhanced the thrills for the public, who would be keen to witness one of their own kind – a mere mortal – actually taking to the air. This in turn would be likely to produce larger audiences and therefore boost both the financial and personal success of these flying shows. Hucks was both a shrewd businessman and an innovative aviator who might well have become the Alan Cobham or Freddie Laker of his day, had it not been for the intervention of the First World War

B.C. Hucks's 1913 season of visits to fenland towns began with his arrival at a grass field adjacent to North Road, Sleaford in the watery afternoon sunlight of Monday 19 May 1913. It was, however, an inauspicious start, due to poor organisation on the part of the local committee and it almost deprived the intrepid aviator of his fee.

A large crowd greeted his appearance, many having queued to enter the field since mid-day but unfortunately, due to the committee's anxiety not to delay the proceedings further, before half the throng had parted with their 1s (5p) entrance fee, the cry 'He's going up!' rippled through the air. Without further ado, those wily country-folk waiting outside the field and queued back over a nearby railway bridge, stood their ground, content to watch the proceedings free of charge!

Disconcerted by this unsporting gesture, the organising committee were naive enough to send round a collecting box but as might be expected, only the paltry sum of 12*s* (60p) was donated. They tried again to prick the consciences of the onlookers remaining for the evening show but received another rebuff when only a derisory 1*s* (5p) was collected. Hucks was far from amused, for even though there were 2,000 people actually in the field, he vowed he would not fly here or anywhere else on these terms, that is to say when only half the potential takings could be realised.

Spurred on, no doubt, by the possibility of both financial loss and ridicule, the committee reassured Mr Hucks about his fee, and entry prior to the following evening's exhibition was better organised. Much to the relief of the committee, paying spectators numbered 1,900 on this occasion and it was noted that of the small number remaining outside the enclosure, the culprits were mainly motor car owners carelessly blocking North Road. It was not, it seems, the prerogative of poor farm workers to be tight-fisted on such occasions, as the local newspaper was also quick to point out!

The programme for the second evening's entertainment opened with stirring renditions by the town band, followed by what became one of Hucks's trademarks – an informal lecture on flying. The flying programme itself comprised three flights, into the first of which Hucks put all the dash, verve, daring and controlled risk for which he was renowned. He swooped upon the crowd, he dived below such sparse contours, trees and buildings as existed and in the style of a true professional, he blew away any bad feeling that might have clouded the event. In a programme designed to show the versatility of the aeroplane he next showed off its height climbing potential, reaching an altitude of 4,500ft and appearing to those country folk like the speck of a bird wheeling about the clear evening sky.

By way of contrast the final flight of the day took Benny Hucks away from the field, out across the flat countryside to the lofty spire of Heckington church, which, it will be remembered, was the scene of James Sadler's momentous balloon journey more than a century earlier. This is also perhaps an ideal juncture at which to put in a good word about the 'trusty Gnôme rotary engine' that powered Hucks's Blériot, by courtesy of that doyen of early aviation writers, R. Dallas Brett. Commenting on the first British Aero Show held at Olympia on 19 March 1909, Brett wrote:

An exhibit which attracted little attention and regarded by many as a freak, was the new [French] Gnôme rotary engine designed by Monsieur Seguin, which appeared for the first time in England. It is impossible to exaggerate the importance of this remarkable invention, which contributed more than any other single factor to the rapid development of practical aviation. The unreliability, coupled with great weight, of contemporary engines drove the pioneers to despair. Monsieur Seguin produced the Gnôme much as a conjurer might produce a rabbit from a hat. At first an astonished audience seemed doubtful if it was real but a trial soon convinced them that here was a reasonably reliable engine which would give an honest 50hp for a weight of 165lbs, which was accessible [to maintain], which took up a minimum of space and was easy to install. It was the answer to their prayer. There was only one serious disadvantage which was that it had to be run on [lubricated by] castor oil, which it consumed in prodigious quantities but that was a small drawback when compared with a gain in performance of at least fifty percent over its nearest rivals. Its immediate success was assured and for the next four years the Gnôme reigned supreme as the aero engine *par excellence*, not only in Europe but all over the world.

There was much speculation riding on the outcome of a sweepstake for the privilege of a passenger flight with Mr Hucks. During the evening of the second day the winner was announced as Mr Doncaster Jr, from the hamlet of Silk Willoughby. For an undisclosed consideration and for reasons best known to himself, Mr Doncaster was induced to part with the singular honour of becoming the first aerial passenger in the fenland region to one Mr Hargreaves of Digby – the young man must have been mad!

Bad weather chose to mar the third and final day of the show, when the wind rose to over 45mph with driving rain from time to time. Despite this, another 1,800 people paid to watch, although that day 400 of these spectators were schoolchildren admitted for just 1*d* each.

The first flight lasted only six minutes and Hucks reported he was buffeted quite severely by the strong gusts. Trying again a short while later, he returned to earth after just eight minutes to be greeted by his worried mechanic who was not at all happy about him flying in such atrocious weather, spectators or not. Conditions did not improve by evening and upon his return to the ground after his third attempt, his face still stinging from driving rain, he reluctantly closed the proceedings. He had intended to fly on to Peterborough that evening but in view of the weather, he postponed that journey until the following week.

As echoes of the band's rendition of 'For he's a jolly good fellow' died away, Mr Norman Snow, chairman of the committee made a closing speech of thanks declaring that they had cleared their expenses from a total of 5,744 paying spectators over the three days. For his personal appearance Benny Hucks received a cheque for £101, which included the £5 fee for taking up a passenger. In addition, as a token of the goodwill of the townspeople of Sleaford, he was presented with the gift of a silver breakfast dish. The winner of the passenger flight did not fly during the exhibition due to the bad weather but he was not to be disappointed as he was taken aloft the following Wednesday, just prior to Hucks's departure for Peterborough next day, Thursday 29 May.

Scarcely was there a breath of air to disturb the atmosphere that afternoon as 3,000 spectators waited expectantly to witness his display in Milton Park, the estate of the Fitzwilliam family on the outskirts of the city of Peterborough. It will be noted that this crowd is considerably smaller in size than that which turned out to see Mr Ewen only a year earlier, possibly reflecting a view that the novelty of aeroplanes was wearing off a bit. Comment to this effect was made in the *Stamford Mercury* newspaper a week later upon Mr Hucks's arrival in that town.

Small crowd or no, Benny Hucks treated his audience to a stirring display of airmanship with four separate flights in his trusty 70hp Blériot monoplane. This was an aircraft he had purchased for £1,100 and flown back from France a year earlier, and similar in design to that in which Louis Blériot first flew across the Channel.

The first of Hucks's flights was described as perhaps the most sensational of the afternoon's programme. After circling the grass landing enclosure he climbed to 2,000ft,

B.C. Hucks and his Blériot monoplane seen at his Milton Park, Peterborough, display on 29 May 1913.

then volplaned (glided, engine off) gracefully almost to ground level. Then, opening the throttle, he skimmed 'like a giant grasshopper' a few feet above the ground. With considerable daring he zoomed up, just over the heads of the spectators, sending them scattering in all directions, for many had thought their last moment had come! In an instant the Blériot was gone from sight, having disappeared behind the stately home, suddenly to reappear at roof-top height from which Hucks side-slipped to earth in a neat landing in the enclosure. The crowd went wild with excitement at the sight of such hair-raising manoeuvres and round after round of applause greeted the intrepid Mr Hucks as he switched off the engine.

On the second flight Hucks took a passenger aloft with him, circling Peterborough Cathedral several times. Mr John Lee, of St Mark's Street, paid the princely sum of £5 for the privilege of becoming the first airborne passenger from that city. He expressed great delight, declaring he wished it could have been much longer.

During the third trip, Mr Hucks demonstrated his electrically controlled bomb dropping technique, releasing two small percussion bombs into the arena, much to the delight of the crowd. Hucks described these bombs as 'filled with a special harmless explosive composition which gives off a most terrifying report and flash without danger.'

For B.C. Hucks, however, time was pressing and refreshed by an overnight stay in Peterborough's finest hotel, his presence was now eagerly awaited in Spalding. So, on yet another clear, sunny, morning, Friday 30 May 1913, he made the fourteen-minute flight at 3,000ft over the flat terrain to provide this south Lincolnshire market town with its first sight of an aeroplane.

A large grass meadow in Stonegate, bordered by the railway line to Holbeach and what are now the residential areas of Alexandra Road and Ayscough Avenue, was selected as the

(Mr Lee Passenger) B Chucks at Milton Park May 24th 13.

With an air of confidence and anticipation Mr John Lee, seated behind Mr Hucks, became the first citizen of Peterborough to take to the air in a powered flying machine. Milton Park, Peterborough, 29 May 1913.

landing ground. A marquee was set up to serve as a hangar for the aeroplane and these arrangements had been made, as was the pattern for most of his displays, by a committee of local businessmen and prominent people, in this case led by Major Barrell. Although Mr Hucks could be relied upon to fly with considerable 'dash', he was, nevertheless, not a man to take unnecessary risks. On this occasion, approaching the landing ground, he found it to be surrounded by high trees, amongst the tallest in the district in fact, which limited the angle of approach. The breeze, too, was quite strong and deciding the risk from wind eddies over the trees was too great, he chose to land in an adjacent field in Clay Lake. Later, a dyke separating the two fields was bridged and his Blériot was manhandled across to the hangar.

On the flying ground itself, admission to which was 1s (5p), every convenience was provided for spectators, including seats and a host of refreshment stalls. The town band played selections of music between flights, and visitors could buy booklets revealing the life and successes of Mr Hucks. For an extra sixpence (2½p) admission the hangar was also opened up to the public where the Blériot could be inspected at close quarters and Mr Hucks was on hand to explain all its intricacies.

Flying, due to begin at 3 p.m., was delayed for an hour by a heavy rain shower and strengthening wind. The sun, however, broke through around 4 p.m. and a buzz of expectation rose from the crowd. Hucks and a mechanic wheeled out the Blériot, the aviator donned his flying overalls and climbed aboard. Without further ado his mechanic swung the propeller and the Blériot took to the sky after a run of less than twenty yards. The

FLYING AT SPALDING ILLUSTRATED.

A montage of pictures showing B.C. Hucks's flying exhibition with a Blériot monoplane in Spalding on 30 May 1913. (*Lincolnshire Free Press*)

first exhibition took the form of several circuits of the local area at 1,000ft, all within sight of the arena. The tall trees still posed a problem for landing and Hucks was obliged to make a very steep approach, with the engine on, to clear them and yet remain within the confines of the field. His subsequent 'round-out' caused a bounce and there was a tense moment of worry before the Blériot finally rolled to a halt, within inches of a barbed wire fence.

During the week preceding the visit it was publicised widely that Mr Hucks would take up a passenger for a £5 fee. The promoters held a raffle for a passenger ticket, which was won by local businessman, Tom A. White. He, however, was a man of robust proportions and considerably above the weight limit stipulated so, greatly disappointed, he sold his winning ticket to Mr Harry Gooch. Mr Gooch was particularly keen on the idea of flying and indeed had even journeyed to Peterborough earlier that very week to try, unsuccessfully, to persuade Mr Hucks to let him fly to Spalding as a passenger.

Hucks's second flight of the display was to have been Harry Gooch's big moment but due to the stiff breeze and the problem with the trees, Hucks postponed both the passenger trip and the bomb dropping demonstration. Instead he took off at 5 p.m. with a view to sorting out the landing problem. After twenty minutes aloft, circling and climbing to become a mere speck at 4,000ft, he decided to land in the field already used for his arrival landing. The only drawback to this was the tedious need to manhandle the aeroplane across a dyke to get back to the hangar enclosure and take-off area. With the wind showing little sign of abating that evening, Hucks announced that flying would cease for the day and he retired to the White Hart Inn for the night.

Dawn of the second day, Saturday, heralded an improvement in flying conditions and a total of four flights were made. Mr Gooch, eager as ever, had to contain his enthusiasm until Hucks, equally determined to fulfil his programme, completed the bomb-dropping routine. At 7.30 p.m., patience rewarded, Harry Gooch, suitably attired in long motoring coat, goggles and with his cap turned around in the time-honoured fashion, stepped forward amid cheers from the crowd.

Even with two persons on board, the little Blériot was airborne in just thirty yards and rose steadily to 2,000ft with the ecstatic passenger waving a handkerchief to the watchers below. After fifteen minutes they landed back at Cley Lake and local newspaper reporters besieged Mr Gooch, plying him with questions about his experience. Bathing in this limelight, he recalled the aerial sights of Spalding and Pinchbeck for the benefit of his audience. He vowed his determination to repeat the experience but no amount of inducement could achieve his aim on this occasion.

The final flight of the day followed the now familiar pattern, concluding with gracious gliding curves, zooms and head-on passes at the crowd, to be met with the same enthusiastic applause from spectators. Such was the success of Hucks's show that the committee, after guaranteeing his fee of course, persuaded him to stay until the following Tuesday. Admission price was reduced to sixpence (2½p) as there were to be only two flights in this display. The Spalding visit was finally brought to a close with Benny Hucks, surrounded by a crowd eager for information, amiably presiding over a lengthy question and answer session about his career and the science of flying.

Now the energetic Hucks pressed on again, this time flying to Nottingham for a two-day show. He then returned to the fenlands on 6 June 1913, as guest of the gentry of Stamford, for his second show in that town. Not for nothing had Benny Hucks earned a reputation as a pilot who would fulfil his obligations and was not afraid to fly in all kinds of weather conditions to do so. Indeed he had been advertised prior to his visit to Peterborough as 'The Rough Weather Flyer'. His journey from Nottingham to Stamford was evidence of his tenacity and flying skill, for it was made in gale conditions and driving rain, which he described later as 'like hot water being thrown in my face'. So bad was visibility that he was barely able to follow the Midland railway line to Stamford and, being driven along at 85mph, passed over the town before realising his error. When he managed to locate the town he was glad to set the Blériot down in Low Park, Burghley, but after seeking directions he restarted the engine himself and flew the short hop to High Park, where a canvas hangar had been erected.

Next day, Saturday 7 June, dawned fine and clear but with a strong, blustery wind. The *Stamford Mercury* was moved to comment that 'perhaps the novelty of aeronautics has worn off, so far as the Stamford district is concerned'. Attendance at the airshow in Burghley Park, it reported, 'was disparagingly small, even though Mr Hucks was performing in aid of various local nursing charities.'

Hucks's show was somewhat muted by the gusty 35mph wind but despite this, his aerobatics were up to his usual high standard and included bomb dropping and a mid-air daylight firework display. By evening the wind had died down and two passengers were taken up as a grand finale. It should be remembered that the cost of a flight, usually at least £5, represented a considerable sum in those days when a working class income was probably only that sum per month. This effectively put the prospect of a flight way beyond the reach of most of the spectators at these shows.

Those being indulged on this occasion were Miss Fanny Aldwinkle, the twelve-year-old daughter of Councillor and Mrs B.W. Aldwinkle, who would not only have the privilege of a flight but, by doing so, became the youngest person to fly. Afterwards, with great aplomb, she described her flight to 2,000ft as 'simply ripping!' The second lucky passenger was Mr W.A. Harvey, a young engineering pupil at the Blackstone & Co. works in Stamford.

Thus, once again, Benny Hucks brought his show to a thrilling climax and left a lasting impression on the youth of Stamford. The relentless pressure of his tight schedule brought him next to the town of Boston.

Writing under the banner headline 'A BIRDMAN AT BOSTON', a reporter from the *Boston Guardian* let his enthusiasm run free and the newspaper devoted a whole page to this, the first ever visit of an aeroplane to the town. Mr Ryan's meadow, on Sleaford Road (the site of Oldrids and Tesco stores in 2007) was to be the venue for the great occasion. At 6 p.m. on Tuesday 10 June 1913, the drone of 'the wonderful eight cylinder Gnôme engine' heralded Benny Hucks's arrival. Majestically, he flew over the town for all to see, circling the Stump, that great tower of St Botolph's church, before 'beating up' the landing ground. In what appeared, however, to be a rare error of judgement, Hucks held off just a bit too high, touched down almost in the middle of the field and was then unable to prevent the Blériot from crashing into the boundary hedge. Willing hands rushed to extricate the aeroplane from its ignominious resting place while the uninjured Hucks ruefully inspected the damage. By way of explanation, he attributed the incident to the fact that he spotted that the enclosure ropes had been placed in a manner that severely restricted his landing run. There was neither sufficient space nor head wind, he went on, to allow the speed to fall away before he hit the hedge. No luxury of wheel brakes in those days!

Without further ado, the offending ropes were repositioned to Mr Hucks's satisfaction but the damage was done and a new propeller was required, together with repairs to some broken control wires. Wheeling the aeroplane to the safety of a makeshift hanger, to keep away inquisitive spectators (some of whom had already scribbled their names on the fuselage and wing linen, and greatly incurring Benny's wrath!), he sent a telegram 'using a page from our reporter's notebook' to the White Hart Inn, Spalding. He had left his spare propeller at this central point for just such an eventuality. Then, being a man unaccustomed to waiting idly by, without further ado he arranged transport to the railway station and caught the next train to London to attend to his business affairs.

In Boston, Wednesday 11 June dawned mistily but with the promise of a fine day to come. Hucks's mechanic, Mr Hamilton, fitted the new propeller and by mid-afternoon the aviator himself had returned from his London office. Duly inspected, the Blériot was wheeled out to the sound of loud applause from several hundred (paying) spectators. Its engine was run up and with a short hop the Blériot became airborne for Hucks to carry out a test flight in his usual dashing style. Back on the ground again, he informed the persistent reporter that the new propeller 'is unbalanced and creating a great deal of vibration.' With his customary courtesy Hucks patiently explained that the damage in the accident had been much more severe that at first thought. An engine cylinder head, numerous wires, the control column and even his seat strap had all sustained damage. The fuselage was twisted slightly and required realignment. However, all this had been repaired overnight and he was confident now that once the propeller was refitted, all would be well once more.

The major event of that first day was to be a bomb dropping display. Hucks's patent electrically operated racks, located in the rear of the fuselage, held up to six bombs. These could be loosed off individually or in salvo – the latter providing a volley of ear-shattering explosions and impressive clouds of smoke.

The *Boston Guardian* report warrants closer examination, providing, as it does, an interesting social comment on the scene at such events. For example, Benny had learned how to extract the odd spot of free advertising for himself and his sponsors but, in fairness, it should be said that he also extracted the maximum publicity for the advancement of the whole concept of aviation too. Constantly harried by the *Guardian* reporter, Hucks let it be known that 'the engine is run only on Pratts Motor Spirit, supplied by the Anglo-American Oil Co Ltd with Mr G W Base the Norwich district manager supervising arrangements very satisfactorily'. This, of course, our ace reporter avidly included in his story.

The flowing style of this reporter also paints a pen picture of that curious meeting of an old with a new era:

> Numerous motor cars and [horse-drawn] carriages were parked in rows … around the arena where, in the summer sunlight their occupants, some sporting parasols and gay bonnets, presented a striking carnival picture.

Hucks made three flights that day, and then the scene was set for a final breathtaking exhibition the following afternoon (Thursday).

Once again it was a gorgeous, sunlit day. Only a moderate crowd gathered in the arena but a much larger throng, on foot and in cars and carriages, was gathered along the Sleaford road. Many people had worked out how to see the show 'fer nowt!' Naturally the organisers were far from pleased but could do little about it and Mr Hucks remarked on behalf of the committee:

> I should have thought that there was enough English sportsmanship in a Boston crowd of *hedge-ticketers* to have recognised by now that this was a genuine exhibition. It does not speak well for them to take advantage of the enterprise of local gentlemen who have arranged the event.

These were fine words but they would carry little weight with Joe Public.

This final day opened with the type of flying which had become Hucks's trademark. He hugged the contours of the ground, out of the crowd's view, and then leapt up over hedgerows and telegraph lines, to zoom and bank right above their heads. After an interval the display was brought to a close with no less than three separate passenger trips.

First to venture aloft was Miss M. Pearson. Self-consciously, she walked beside Mr Hucks across the arena to the waiting Blériot, to a rousing rendition of *See The Conquering Hero Comes* from the local brass band. Upon her rather breathless return to terra firma she, somewhat less demurely, exclaimed, 'Ooh! It's just ripping!' and about the view, 'It was just divine! It was awfully funny when you looked over the side.' Once again we are indebted to our gallant reporter for the accuracy of the lady's remarks. Mr G. Holland, a leading light of the organising committee, purchased the winning sweepstake ticket and thus became the second passenger. He, too, was roundly cheered upon landing and could only emulate Miss Pearson's 'Ripping!' as his verdict on the flight, which had included an aerial view of his own motor car garage. A further interval gave an opportunity to squeeze in an aerial firework display, after which came the final flight of the day. Local businessman Mr A.E. Beulah was the passenger and upon his return to earth his sole comment was a taciturn 'Very fine!' Everyone considered it had been a most splendid display and Benny Hucks could, justifiably, be very pleased with his efforts as he retired to the comfort of the White Hart Hotel, in the shadow of Boston Stump.

With Louth and Lincoln in June and July the next venues on his relentless aerial progress around the country, the fenland region took a last look at this famous flyer at Skegness:

B C HUCKS, The Famous Airman, will give thrilling Demonstrations of Flying at The CAMP GROUND (Cricket Field Lane, near Station), on August 19, 20, 21, 1913.

Thus ran the bills circulating in the seaside town of Skegness, while the local newspaper summed up his visit as follows:

A huge crowd gazed skywards at what appeared to be a great dragonfly careering about in eccentric circles over Lumley Road. Shops emptied and people ran into the streets to catch a glimpse of the birdman. His exhibition was the talk of the town for months afterwards.

Benny Hucks had that sort of effect on people.

It also appears this was not his first exhibition at Skegness. The *Stamford Mercury* carried an advertisement in its issue of 20 June 1913, for a display by Mr Hucks at Skegness from 26–28 June. In that same newspaper two weeks later, a report read:

Brilliant flying on 26th. Mr Hucks made a trial flight in the morning, over the town and along the seashore. It served to whet the appetite and a large number of people journeyed to the Camp Ground to witness the display in the afternoon and evening.

Skegness in the height of the summer holiday season presented a business opportunity too good to miss.

Considering his business commitments and the punishing flying schedule he set himself, it is hardly surprising to learn that during this period of his career Hucks suffered a serious attack of pleurisy. On the outbreak of war in August 1914 he volunteered at once for active service with the RFC. Hucks did a considerable amount of operational flying in France before the effects of his lung complaint caught up with him again. His health gave way and he was invalided back to England.

Regaining his health once more, he undertook aeroplane test flying work, for which he was of course eminently suited. Subsequently he was attached by the RFC to the

Advertisement for B.C. Hucks's flying exhibition
in Skegness in August 1913.

Aeroplane Manufacturing Co. Ltd (Airco) as a test pilot. Now promoted to captain, he
spent the remainder of the war test flying the designs of Geoffrey de Havilland and in
this capacity Hucks made a huge contribution to the war effort. In addition to his work
in helping to produce a string of fine DH production aircraft, such as the DH9 series, he
will be remembered also as the force behind the creation of the 'Hucks starter'. This was
a curious but ingenious adaptation of a Model–T Ford, used to ease the starting of large
aeroplane engines before the days of the inertial starter. An example of this weird but
effective contraption can be seen in the Shuttleworth collection at Old Warden.

Sadly, Bentfield Hucks died on 6 November 1918. He succumbed to pneumonia after
contracting influenza during the pandemic sweeping the country at that time. C.G. Grey,
that highly respected editor of *The Aeroplane* magazine, wrote of Hucks:

[He] was one of the most likeable of men. Though very certain of the correctness of his
own opinions, and full of confidence in his ability and judgement as a pilot – as his long and
successful flying career entitled him to be – he never thrust his theories down the throats of
others and he never 'swanked' about his flying. A pilot of superlative skill, a mechanic of more
than ordinary ability, a sound businessman and a good sportsman, Benny Hucks would have
gone far in the future developments of commercial aeronautics had he lived.

The whole aviation world mourned his passing.

Many airmen aspired to Hucks's crown but few managed to get close. Lincoln, for
example, was promised 'The most wonderful exhibition of flying England has ever

THE MOST WONDERFUL EXHIBITION OF
FLYING
ENGLAND HAS EVER SEEN.
THE DARING HENDON AIRMAN,
MR. E. R. WHITEHOUSE,
on his Handley Page Monoplane at
NETTLEHAM ROAD, LINCOLN,
ON FRIDAY, JUNE 27, AND ROSE DAY, SATURDAY, JUNE 28
FROM 3 P.M. UNTIL DUSK.
BOMB DROPPING, PASSENGER CARRYING, FANCY FLYING, AND
FLYING IN THE DUSK GUIDED BY ROCKETS AND FLARES.
BAND and REFRESHMENTS on the FLYING GROUND.
GRAND FIREWORK DISPLAY (EACH EVENING).
The Best part of the EXHIBITION can only be viewed from WITHIN the Ground.
POPULAR PRICES OF ADMISSION.
Organised by the Scientific and Instructive Aviation Co., Ltd., Town Hall Chambers, St.
Albans. Local Representatives: Messrs. GILBERT & SONS. Ltd. Pelham-street, Lincoln

Advertisement for a flying display by 'the daring Hendon airman' E. Ronald Whitehouse, at Nettleham Road, Lincoln, on 27 June 1913.

seen!' but it was a claim that E. Ronald Whitehouse – billed as 'the daring Hendon Aviator' – had to struggle hard to back up.

Mr Whitehouse's flying exhibition on Friday 27 June 1913 presaged the city's annual Rose Day weekend of festivities but his frail Handley Page 5 'E' monoplane was no match for the strong wind that whipped across the airfield on Nettleham Road all day. By evening, though, the wind had dropped sufficiently to treat the large gathering of spectators to what the *Lincolnshire Echo* described as 'a magnificent flying exhibition.'

Ronald made several short flights to begin with. One of these involved taking up a passenger – but not actually coming back with him! This is not really as alarming as it may have looked to the spectators on Nettleham Road. When they got up in the air and flying around the city, Whitehouse found the aeroplane very difficult to control in the increasing wind. Discretion was the better part of valour so he landed on South Common, a couple of miles away, unloaded his passenger and returned to Nettleham Road alone. The poor passenger was left to find his own way back as best he could.

The greatest thrill for the crowd was reserved until last. With the light fading as Ronald rounded off his lecture to the crowd on the workings of his aeroplane, he donned his hat and goggles, climbed into the monoplane and took off into a stiff breeze and the gathering darkness. In the words of the *Lincolnshire Echo*, 'He treated the spectators to an exciting flight during which he reached an altitude of 3,000 feet and disappeared into the gloaming to the north-east of the city.'

If it were that dark it seems most unlikely the crowd would see very much from the ground so it may well have been seen in the reporter's vivid imagination. Ten minutes

E. (Evelyn) Ronald Whitehouse, as he
appears in his pilot licence photograph.
(RAFM)

passed by with no sign of the dashing young aviator. As anxiety murmured through the
ranks, rockets were fired into the air and acetylene flares lit to guide him back to the field.
Another ten minutes passed before the clatter of the 50hp Gnôme engine released the
mounting tension as the Handley Page dipped like a moth out of the gloom to make a
perfect landing and prove to everyone that it was indeed possible to fly in the dark.

Unfortunately, the next morning (Rose Day) a pre-flight inspection of the Gnôme
rotary engine revealed a cracked cylinder. Whitehouse scoured the country by telegram
for a spare part but without success. There was no option but to order the part from
France and cancel the remainder of the Lincoln programme. By the end of the following
week the repair was completed enabling Ronald Whitehouse to fly off to his next
engagement in Hull.

Another year would elapse before the clatter of a rotary engine could be heard once
more in the region. While in 1914 there were no more than 100 qualified British pilots
actively engaged in civil flying, Hucks still enjoyed deserved acclaim as the foremost
among these aviators. Another airman, Frank Goodden, was also considered by many to
rate a place in the top half dozen exhibition pilots of that year. A graduate of William
Ewen's flying school at Hendon in June 1913 (certificate No.506) Goodden had come
into the powered flying age by way of parachuting from balloons, much in the manner
as the events of the previous chapter. He was actually involved with the first descent by
parachute from an aeroplane in England, in May 1914, although he did not make that
particular jump himself.

Frank Goodden, one of those magnificent airmen who enthralled the British public in the years before the First World War. (Maurice Buck)

It was on Thursday 26 June 1914 that, in the words of the *Peterborough Advertiser*, 'great popular interest was taken at Peterborough in the spectacular feat of looping-the-loop as performed repeatedly by Mr Frank Goodden, the brilliant young aviator'. In the struggle to make a name in the flying game, looping-the-loop was the most sensational feat currently exploited to excite a crowd and keep the gate-money flowing. Feats of navigation, however, were certainly not a skill at which Goodden excelled. Trying to find his way by air from Luton to Peterborough that morning he lost his bearings. He had been following the railway line north from Luton but inadvertently turned left at Sandy and after landing to enquire as to his whereabouts, found he was following the line towards Oxford instead of Peterborough!

Soon after 4 p.m., though, the Morane Saulnier monoplane, powered by an 80hp Gnôme rotary engine, was wheeled out of its temporary hangar to the strains of bagpipe music. This was the aeroplane with which Gustav Hamel until recently conducted his own immaculate flying exhibitions. Hamel had gone missing, presumed lost in the Channel, in May 1914 while ferrying another machine back to England. Frank Goodden, Hamel's flying partner, acquired the Morane, now painted black in mourning for his dead companion. In taking over the mantle of Hamel he retained the services of mechanic Paul Goudré and the piper Donald Fraser.

Now a resounding cheer greeted the appearance of Goodden. Climbing into the cockpit he was assisted with the full body harness by Goudré, who then swung the propeller to start the engine. In almost complete silence the crowd at the agricultural show ground in

Frank Goodden, as he appears in his
pilot licence photograph.
(RAFM)

Eastfield watched as the aeroplane 'rose like a bird'. Circling the field Goodden took the
Morane up to 1,000ft then, to gasps of astonishment from spectators he flew into wind,
towards the direction of Eye and pulled up into the vertical, over onto its back, with the
engine idling now, then with a swoop earthwards his loop was complete. Four more times
this manoeuvre was repeated, interspersed with rolls, steep banking turns, flinging the
little Morane about on a sixpence to the obvious delight of those below.

Even by 1914, despite aeroplane exhibitions having become commonplace in the
popular culture of the times, pilots were still regarded with awe. The *Advertiser* reinforced
this image thus:

> Goodden has no nerves. He went up laughing; he came down smiling, jumped from the
> cockpit and lit a cigarette. He submitted to being besieged by young ladies clamouring for his
> autograph. If it is a clever, amazingly daring thing he does, HE apparently is the last person
> to think of it thus.

A cool dude!

Goodden repeated his show later that evening and took a few passengers up in the back
seat. Among these lucky individuals was Mr A.A. Turner of Fletton who won his seat as
the result of a lottery for a free flight, while another was the adventurous Mr Lee, who it
will be remembered was the first passenger from the city to fly when Hucks visited the
city in May 1913. In a tightly packed schedule it was Goodden's intention to fly to his next
venue in north Wales the following day, weather – and his navigation – permitting!

Above and below: Frank Goodden landing his Morane monoplane after a stirring aerobatic display in Peterborough on 25 June 1914. (Maurice Buck)

When war came Frank Goodden, like many of his contemporaries, volunteered his services to the aero industry. Commissioned into the RFC he became chief test pilot at the government aircraft factory in Farnborough and rose to the rank of major (squadron leader). He was involved with flight-testing many well-known types such as the BE2, RE8 and SE5 and indeed the latter highly successful fighter was originally conceived from design ideas put forward by Goodden. He ferried A4562, the second prototype of the latter, to France on Christmas Eve 1916 for its operational trials with 60 Squadron, whose pilots evaluated it against their Nieuport fighters. Back in England in January 1917, Frank continued to test A4562 until on 28 January it suffered a fatigue failure in the lower wing spar while he was halfway through a loop – ironically the very speciality on which he had built his reputation. The wings collapsed and twenty-six-year-old Frank Widdenham Goodden was killed in the crash. It must be said he did not die in vain because a thorough investigation of the wreckage identified a fundamental design flaw in the wing spar. This caused fatigue to accumulate when the wings were subjected to the sort of loads encountered in inverted and other negative-G manoeuvres. The design was rapidly changed and the SE5 went on to prove itself in battle. There were many glowing tributes to Goodden:

> Tall, handsome, technically knowledgeable and possessing that charm and vividness of spirit which lends itself to leadership, his passing left a blank. Nobody had the same extensive testing experience.

Even while Europe was already sliding down the road to war, the *Daily Mail* extended its sponsorship of aviation to include a separate tour by seaplanes. As part of this tour F.P. (Freddie) Raynham, a test pilot for A.V. Roe and another noted display pilot of the day (RAeC certificate No.85), flew an Avro 504 Waterplane owned by the *Daily Mail* to Hunstanton, at the north-eastern extremity of The Wash. The date was Tuesday 8 July 1914, less than a month before the outbreak of the First World War, but sweltering in the summer heat the public seemed pretty blasé about events happening 'over there' in Europe.

This two-seater biplane, prototype and precursor of that long Avro 504 line, powered by an 80hp Gnôme Monosoupape rotary engine, was fitted with floats in place of its normal wheeled undercarriage. Emblazoned on each side of the fuselage was the legend *Daily Mail* and on the fin/rudder Avro biplane. On the Monday a hangar was erected on the green in front of Beach Terrace and it was here, after Freddie had flown in, that Mr Ridgeway, the tour manager, and four mechanics worked to fit new float struts to replace those damaged when the Avro had a mishap during a previous exhibition on the Isle of Man. The struts obstinately refused to fit into their sockets so much time was wasted rectifying the problem. By 4 p.m. everything was fixed; the Avro was manhandled down to the water and Raynham took off for a short test flight. Over 12,000 people, spread out all along the promenade and cliff top, watched his arrival and initial display and they gave him a rousing reception with clapping and cheering when he cut the engine on his return.

Now the real business got under way. The first paying passenger was William Bates of Homerton, Cambridge, a holidaymaker staying in Hunstanton. Next up was Mr F.J. Blakey who was staying in Burnham Deepdale and claimed he had really only come to Hunstanton to see a doctor! After his flight he said, 'When I went up I was feeling quite seedy but the flight was a real tonic to me so I shall go straight home and save the doctor's fee!'

That evening a splendid dinner, laid on by the Advancement Association, was held in the Sandringham Hotel in honour of Mr Raynham and his *Daily Mail* team. It was

THE DAILY MAIL AEROPLANE TOUR, 1914.
MR. F. P. RAYNHAM AND "AVRO" WATERPLANE, WITH 80 H.P. GNOME MONOSOUPAPE ENGINE.

A souvenir card showing Freddie Raynham and the prototype Avro 504 during the *Daily Mail* Waterplane tour of 1914. (Ray Wilson)

attended by many local dignitaries including the chairman of the Urban District Council, J. Bowman Esq., and J.T. Halliday representing the British Petrol Company.

Thursday (9th) dawned with the sun shining brilliantly again and the sea was as calm as a millpond with just a hint of a gentle breeze from across the shimmering expanse of The Wash. At midday, before an expectant but somewhat smaller, gathering of locals and holidaymakers, Raynham made his first flight of the day, which lasted all of seven minutes. Shortly after this he took his first passenger aloft, a young (but unknown) female holidaymaker. This flight, too, was short – a mere five minutes – but the crowd was buzzing with the excitement of seeing ordinary people flying. No sooner had the young lady returned to shore than Raynham was off again, this time with local resident Mr H.A. Durrant as the lucky passenger. Upon his return he decided to call a halt for lunch.

Flying resumed at 3.15 p.m. with Raynham making a short solo flight, no doubt to check that all was well with his aeroplane before continuing his passenger trips. He performed some aerobatics well out over the sea and above the King's Lynn fishing fleet busy with its nets, then returned to pick up Miss Gladys Angell. Miss Angell owed her six-minute sky ride to being the lucky winner of the *Daily Mail* raffle for a free flight coupon. The prize was actually won by a Mr J.M. Head of Old Hunstanton but when his name was called he did not step forward to collect his prize so the raffle was re-drawn and Miss Angell's name came out of the hat. She wasted no time and was off like a shot, wading out to the boat that took her out to the seaplane. Next up was Mrs C.E. Grey of the Golf Links Hotel in Hunstanton, and the final paying passenger was Mr W.H. Johnson.

Freddie Raynham, as he appears in his pilot licence photo. (RAFM)

He had a delightful ten minutes in the air and was heard to declare as he stepped back on dry land:

> It were just like sitting in a comfortable arm chair and if I could only have taken my pipe with me, I would have been content to have been taken a long way in that aeroplane.

At 5.30 p.m. Mr Raynham brought the event to a close. With his mechanic firmly strapped into the aeroplane he performed a lively display of 'pretty evolutions' above the pier, and with a final swoop along the sea front Freddie Raynham was lost to sight heading east to follow the north Norfolk coast round to Cromer, his next venue.

His programme at Cromer followed the same pattern as Hunstanton but no flying took place on Friday 10 due to the coast being shrouded by sea fog all day. The canvas hangar was erected on the beach on the town side of the east groyne and there was a constant flow of people coming to look at the notice board to see if there was any chance of a flying display. By mid-afternoon the only thing that was clear was that the fog would not clear. It took until 2 p.m. the next day, Saturday 11, for the sky to clear and the sun to return. Five lucky passengers flew with Freddie that day: Mr A. Collison, Mr A.G. Russell, winner of the free flight raffle, John Smith, Miss May Little and Mr J.M. Head. By 4 p.m. on Monday 13 July more thick sea fog cleared sufficiently for a final spirited flying display by Mr Raynham, followed by the only passenger trip when the winner of the day's free flight, Mr P.J. Cook of North Walsham, closed the show. Freddie Raynham left Cromer by car shortly after 5 p.m. and the Avro was dismantled to follow him to Brooklands by rail. At a council meeting later that evening Mr Ridgeway attended to accept on Raynham's behalf a silver cigarette case as a token of appreciation from the residents and visitors of Cromer.

Two days after war was declared, this Avro 504 suffered engine failure while Freddie was delivering it into war service as an impressed aircraft, and it was damaged beyond repair. Freddie Raynham's contribution to the war effort was, like many of his contemporaries, in the field of aeroplane testing. He worked as a freelance civilian test pilot variously with Martinsyde Ltd, A.V. Roe and the Bristol Aeroplane Co, flying a wide variety of fighter and bomber types. In 1919 he and his navigator, Major C.W.F. Morgan, made a bid for fame when they entered the field of starters hoping to be the first to fly the Atlantic. They failed when their Martinsyde *Raymor* aircraft crashed on take-off in Newfoundland, fortunately without seriously injuring its crew, their effort being overshadowed by the success of Alcock and Brown.

The curtain came down on the region's pre-First World War airshows, with the final act taking place on Tuesday 21 July, at the Horse Show held in glorious summer weather on the recreation ground in the fenland town of March. Country life in England still seemed to be sublimely unaware of the gravity of international events or, at best, apathetic towards them.

Apathy towards aeronautics was displayed by some elements of the press, too. For example, Marcus Manton's two hour exhibition that evening warranted a mere eight lines at the foot of a column in the *Peterborough Standard*, sandwiched in fact between a long discourse about horses and an advert for constipation pills! Fortunately the *Ely Standard* did a much better job, providing us with a very detailed account of this, the first time an aeroplane had visited March. The reporter waxed lyrical:

> Vast crowds assembled in the evening to witness the exhibition of flying by Mr Marcus Manton, the famous 'boy aviator' and the tremendous amount of interest that the event created is regarded as ample justification of the policy of the Committee in providing it. The evening amusement problem has always been a very vexed one for the Society and the Committee has not always succeeded in bringing about a satisfactory result. That an aviator should be engaged to give flights was an ambitious stroke and one deserving of the utmost success.

During the afternoon Manton's Blériot monoplane, powered by a 50hp Gnôme rotary engine, was open to public view for a small entrance charge. Our ace reporter went on to describe in minute detail the construction and characteristics of the aeroplane and had quite obviously paid great attention to Mr Manton's introductory talk, mentioning, for example, 'The two bladders in front of the pilot's seat, which excited a good deal of curiosity, are for keeping an air pressure in the petrol tank during upside down flying.'

Marcus Manton, as he appears in his
pilot licence photograph. (RAFM)

Flying commenced at 6 p.m. watched by a large crowd on the field and, inevitably, even
more watching for free in the lanes around it.

> There were in all five flights and the aviator reserved the looping the loop until the last,
> which was the veritable tour de force of the evening. Mr Manton here attained his greatest
> altitude and for a time his machine appeared to be little more than a speck flying well nigh,
> it appeared, through the clouds. The climax was now to come. Suddenly the monoplane
> made a huge dive, almost simultaneously it tilted right over and flew for a time in the
> reverse position. These aerial acrobatics were repeated again and again, the intrepid aviator
> exhibiting a superb control over his machine. Compared with 'looping' orthodox flying must
> be accounted quite prosaic.

Questioned later about his flying, Marcus said his loops were performed at 2,000ft altitude
and his speed was 55mph flying horizontally and 100mph in the dives. He mentioned he
experienced some discomfort during the flights and explained:

I have to wear numerous leathers, of course, for purposes of protection and the heat and friction tear the skin almost completely off my arms. I had just got better in this respect since my last engagement but now they are quite raw again.

A disciple of Bennie Hucks, Marcus Manton began his exhibition career travelling with the great man, the two of them often performing dual synchronised aerobatics in a pair of Blériots. During 1914 Hucks split his operation into two shows, each touring the country independently, hence the appearance of Marcus Manton and his Blériot monoplane at March on this occasion. However, Manton's show was the last of an era in the region for, just two weeks later, on 4 August 1914, the reality of war finally arrived. Joining that illustrious band of pre-war exhibition pilots who were to make such a valuable contribution as test pilots to the British aeroplane industry in the First World War, he spent most of his time as a civilian test pilot with J. Samuel White Ltd, a builder of seaplanes on the Isle of Wight whose aircraft designs were marketed under the name of 'Wight'. Manton survived those experiences and in 1919 joined the Airco company as manager of its Hounslow airfield.

Upon declaration of war, the Home Office immediately issued an order prohibiting all cross-country flying by civilian pilots and drew a veil of secrecy over all aeronautical activity.

Civil flying came to an end – at least for another four and a half years.

CHAPTER FOUR

Barnstormers

Tragedy projected Donald Hastings Sadler into the public limelight, which was something the routine of his wartime flying career had been unable to achieve.

He was born in London in 1899, and after leaving school Donald followed his father's footsteps into engineering. Joining the RFC in 1917, he qualified as a pilot with the rank of lieutenant and spent the remainder of the First World War as a flying instructor at Croydon aerodrome. It was here that his love of flying took root before, with many hundreds of flying hours to his credit, he was discharged in April 1919. One of thousands of pilots thrown suddenly into civilian life at the end of hostilities, Sadler fell back on his skill and experience as a pilot to gain employment. This experience and enthusiasm stood him in good stead as he was able to find a post on the flying staff of A.V. Roe Ltd, the aircraft manufacturing company, with whom he stayed for about a year.

In 1920, however, with the aircraft industry in the doldrums and A.V. Roe cutting back its joyriding activity, his thoughts turned towards other employment in the exciting new world of commercial flying. With 1,000 flying hours in his logbook and still only twenty-one years of age, he joined Samuel Summerfield & Co. of Melton Mowbray, one of many small air transport businesses springing up across the country. Sam Summerfield's aim was to bring aviation to the people by what became known popularly as 'barnstorming'. This was the itinerant process of moving aeroplanes, pilots and flying shows from town to town, putting on aerobatic displays to attract crowds from which a number of people would be persuaded to fly as passengers for a fee.

The scene is high summer 1920. Summerfield & Co. was touring Midland towns with an Avro 504K, G-EADR, offering business and, primarily, pleasure flights to the public. A week's flying at Luton was followed by two weeks in Northampton, then on to Peterborough, with Lincoln to follow.

Donald Sadler was pilot for this group, whose other members comprised Samuel Summerfield, proprietor and venue organiser, his brother Albert, acting as ground manager, with ground crew of Frank Gordon (airframe rigger) and Joe Kitchener (engine fitter). It was a nomadic life, with a steady flow of flying, plenty of fresh air, new faces and places; this lifestyle is probably what appealed most to Donald.

He would have been familiar with his mount, too, since it was a type on which he had done much of his service flying. Powered by a 120hp Le Rhône rotary engine and built originally as a tandem two-seat trainer for the RFC, in which it served as D6245, this Avro 504 was bought by Summerfield's in March 1920 from another civilian owner, the Bournemouth Aviation Company. In its civilian guise G-EADR was modified to

Summerfield's ill-fated Avro 504K, G-EADR, at Werrington in June 1920, with pilot Donald Sadler third from right. Also in the picture are airframe rigger, Frank Gordon, first left and engine fitter, Joe Kitchener, second left, with Sam and Albert Summerfield third left and far right.
(John Knight)

accommodate two passengers in the rear cockpit in addition to the single pilot's cockpit in front.

Peterborough basked in warm summer sunshine and fêtes, sports and galas all completed with the flying tour. Few people seemed interested in pleasure flights during that afternoon of 24 June 1920 but at 1 guinea (£1 05) a trip, flying was not for the pocket of your average working man. Financially, as well as socially, times had changed little since those far-off ballooning days.

Fate played a curious hand, too, in events that afternoon.

Two passengers, Mr Weber and Mr Gibbons, had just landed but remained seated in the rear cockpit. They expressed a wish to go up again because they wanted to fly over a particular place. Waiting patiently nearby were two other prospective passengers, Charles Guest and Philip Runquest. The latter was lodging with Mr Gibbons and therefore begged his friends to give up their place as he was somewhat pressed for time. Gibbons and Weber were in no particular hurry so they agreed and gave up their seats to Guest and Runquest. This unselfish gesture undoubtedly saved their lives, for the aeroplane crashed during its very next flight, killing Donald Sadler and his two passengers. A *Spalding Guardian* reporter who was passing the field at the time of the disaster wrote:

I happened to be an eyewitness of the terrible disaster at Walton on Thursday. I was walking along the Marholm Road just opposite the aerodrome. Soon after half-past three the aeroplane rose, made a short trip and returned but did not descend. A series of brilliant displays of clever pilotage followed. Finally it descended and I saw it resting in the field for about a quarter of an hour. That was the last complete flight it made.

Now it took off again and went for a rather lengthy trip over Peterborough. This was its fatal trip. As it approached the aerodrome on its return journey, the machine was put through one of the turning over or spiral flights downwards and it was soon afterwards that the tragedy occurred. Spiralling several times towards the earth, it rose again to a height, roughly speaking, of about 600 feet. It sailed along gaily, when suddenly what seemed to be several longish sheets of paper or cloth broke away from the aeroplane. For a brief second or so the aeroplane, in its broken shape, was seen as it began its headlong rush downwards but it swiftly became a long black and fiery line. When it struck the earth it was with a simultaneous loud explosion, while flames shot up to a height of twenty feet, continuing to burn fiercely.

A young man, with his bicycle, stood nearby me watching the evolutions and when the machine began to fall I heard him gasp. When it hit the ground, I heard, without looking round, an exclamation: 'My God!' I turned to some ladies near me to see if all was well with them but they pointed to the young man, who had fallen in a faint onto the grass. I moved to help him but he scrambled to his feet, mounted his cycle and rode off as fast as he could pedal. No one spoke for a minute or two. We knew we had witnessed one of the most shocking tragedies in the history of Peterborough. Yet I knew the sun was shining brightly, that birds were singing, that fields were bright with flowers and hedge-rows ablaze with dog roses but there, silently now, lay the aeroplane and charred bodies of those who but a few minutes ago had been in the heyday of a summer's day holiday. Such things happened daily in the war and many more besides, but they took place amid the roar, din and hell of warfare. This took place in a peaceful and bright setting.

The aeroplane crashed to earth 200 yards from the Cock Inn in Werrington village, on a site now occupied by a small trading estate, between the main railway line and the A15 road. A detailed examination of evidence could not establish a precise cause for the aircraft breaking up.

Despite this misfortune Samuel Summerfield remained in the barnstorming business for many years, but he seems not to have had the best of fortune when it comes to accidents, since after EADR he lost two other Avro 504s in mishaps before he decided to give up being in business on his own account. EAEB crashed at Norwich in 1921 and EASA at Doncaster in 1922. Public records show him to have obtained a 'B' Licence (No.683) in 1922 and his name appears in a list of pilots employed in 1929 by the Wolverhampton Aviation Company – of which more later. Furthermore a 'Capt Summerfield' is also recorded as a pilot of one of Cornwall Aviation Company's all-red 504s when that company toured the Cleethorpes (Lincolnshire) area around 1927. This migration of men and aeroplanes can be seen as an indication of the close-knit nature of the British barnstorming fraternity in those days.

Up to and including this incident in 1920, in more than a century of civil flying only four civilians had died in the region as a result of an accident in the air but, curiously enough, all of them were killed in Peterborough.

Stirring tales of air battles over the Western Front during the First World War fired public imagination but civilian flying had been prohibited since August 1914 when war was declared. Although hostilities ceased in November 1918, it was found that, legally speaking, the Air Ministry had no authority to regulate civil flying unless new legislation

Pilot Donald Sadler with his airframe rigger and engine fitter, in relaxed mood in front of Avro 504K, G-EADR, on the fateful day of 24 June 1920. (*Lincolnshire Free Press*)

was created for that purpose by Parliament. To fulfil this requirement the Air Navigation Bill came on to the statute book on 1 April 1919 and the Air Ministry actually lifted the ban the following month. The government considered it necessary to apply a regulatory regime to civil flying in the post-war era because, as members of the Air Council put it:

> There is no desire to impose upon civil flying any restrictions that would tend to prevent people evolving new types or doing anything they think would advance flying but it is necessary to have the power to secure the safety of the public. There will be control and inspection procedures for all aircraft plying for hire and reward and certification will be done daily by persons appointed by the manufacturers with the approval of the Air Ministry.

Thus was the way opened for a lucky small percentage of the 'surplus' airmen – such as Donald Sadler – to earn a living, fulfil their craving for flying and satisfy the public's curiosity. And where better to meet the public at a time when it was most likely to have money in its pocket, than at a seaside holiday venue?

Despite inclement weather in mid-August 1919, Hunstanton was still crowded with visitors, some of whom were looking for a thrill that was a bit out of the ordinary!

Established at Hardwick near the university city, Cambridge School of Flying was, as its name implies, formed for the purpose of providing some of the first post-war civilian flying tuition. In order to augment its income further, the school detached one of its Avro 504Ks, G-EAHL (ex-RAF E4118) to Hunstanton for joyriding, under the supervision of

Passenger-Flying at Hunstanton

by the

Cambridge School of Flying.

Short & Long Flights, Avro Machines,

from 30/-.

FLYING THIS WEEK. | NEXT TO THE PIER.

Chief Pilot, Capt. Birkbeck, D.F.C.

(By kind permission of the Hunstanton Town Council, and under Authority from The Civil Aviation Control Board, Air Ministry).

Barnstorming comes to Hunstanton in August 1919. (*Lynn News & County Press*)

pilot, Captain Robert Birkbeck DFC, a First World War ace who had flown Nieuport fighters in action with No.1 Squadron.

By kind permission of the local council and subject to tides, the landing site for this venture was to be a 300 yard strip of firm beach south of the breakwater, between the pier and Gas Works Road. Press advertisements and posters offered holidaymakers a chance to take 'short or long flights, from 30 shillings [£1.50]'. For an extra 10s (50p), stunts such as looping the loop, spins etc were available to the more adventurous passenger.

Delayed for a few days by bad weather, the Avro duly arrived at Hunstanton on Saturday 23 August 1919 and by Sunday evening had made several passenger flights. Monday's bad weather prevented flying but Tuesday cleared up and more trips were made. Wednesday and Thursday were also non-flying days and the rain persisted right through until the next Monday (1 September), keeping the aeroplane firmly inside its temporary hangar at Sedgeford, on the site of the former First World War airfield. It is recorded that passenger rides 'last about ten minutes in the air.' One obviously satisfied customer commented, 'I felt no shocks or unpleasant nerve attacks when I took my flight. I was one of the last to go up today and I travelled to Sedgeford aerodrome where the plane is lodged and returned to Hunstanton by road.'

Only a few more passengers were able to go aloft on 1–2 September, before a combination of declining numbers of holidaymakers and quality of weather made it sensible to return the Avro to its base in Cambridge. As the year moved towards autumn these flying enterprises were facing declining revenues and had to work hard to keep the money flowing in. With this in mind the proprietor of Cambridge School of Flying, Capt. J. Lee-Jones, secured work on the other side of The Wash at Skegness in late September 1919.

The *Skegness, Mablethorpe & Alford News* informed its readers that Capt. Lee-Jones anticipated his first aeroplane, Avro 504K, G-EAHL, flown by Capt. Kelley and Capt. Edward E. Fresson, would arrive in Skegness on Friday 19 September for a week of flying. From his point of view it made good business sense to invite local councillors to take free flights as soon as the machine arrived, although it is hard to judge whether the following comment made in the newspaper about that 'perk' was meant to be sarcastic or not!

> It is pleasing to know that at last the representatives of the town are to receive some recognition for their services to the town.

Another reason for this appearance of Cambridge Flying Services in Skegness was because the local council had, for reasons not explained, been unable to carry through a prior

Avro 504, G-EAHL, flown by Capt. Robert Birkbeck at Hunstanton in August 1919. (A.J. Jackson collection)

arrangement with yet another barnstorming organisation called the Navarro Flying Co. The council had to find a replacement quickly and Capt. Lee-Jones willingly stepped into the breach because it was also keen to finalise arrangements for next summer's flying attraction at the resort.

Prices for flights were the same as those in Hunstanton but 30s in the post-war austerity of 1919 would represent a tidy sum to most people in those days – being at least equivalent to perhaps £30 to £40 in 2007 values. Although it was observed by some that 'it was a great reduction upon the £5 that was paid in 1913 (to Bennie Hucks etc), it would soon become clear to flying companies that they would not find enough customers to make it pay unless the price per trip was dramatically reduced. Prices quoted later on in this chapter seem to confirm this view.

Since flying took place at low tide from the beach near the 'Figure Of Eight', the approximate times when this would occur were printed in the local newspaper for the convenience of holidaymakers. The aeroplanes attracted huge crowds every day – but mostly to watch the fortunate paying passengers. Councillor Haley and Mr Frearson, clerk to the council, were the first to take advantage of a free flight and declared they were highly delighted with the aerial tour of the town. From his vantage point at 3,000ft Councillor Haley said:

> The pier looked like a plank laid along the sands to the pier head, which appeared like a small hut with a circle around it. The white circle was the waves dashing against the piles. The sensation of gliding through the air was beautiful, whilst climbing upwards was hardly discernable.

Business was in fact brisk enough to make it worthwhile for Capt. Kelley to go back to Cambridge and collect a second aeroplane, DH 6, G-EALT, so that the two airmen could cope with the number of holidaymakers wanting to experience the thrill of flight in the few days remaining.

There was an interesting sidelight on the flying visit when, in his end of season report to the council the following month, the Skegness Sands Inspector said he had had trouble from stallholders trading along the beach about the aeroplanes being able to fly on a Sunday while regulations prohibited them from opening for business. The inspector was aware that Cambridge Flying had applied to operate from the foreshore site next year, and in his opinion a clearer arrangement must be made in order to placate the stallholders. Flying monopolised the beach, which he felt was not a proper place for it as it was potentially dangerous and perhaps the flying company ought to hire a field away from the beach instead.

Although unemployment was to rise with each passing year, the thrill of barnstorming grew in public popularity, as the austerity of the war and its aftermath receded. In order to make something of a living and cover costs, flying companies were obliged to streamline their operations and maximise time in the air. Furthermore, to attract sufficient passengers, fares had to be reduced to a level that would induce the 'ordinary' person to part with his money in these hard times. It was just not sufficient to entertain spectators – you had to persuade some of them to part with their money!

From a pound or guinea (£1.05) fare in 1920, it was possible by 1923 to fly for 5s (25p) – still a tidy sum in those days but now much more attainable. Equally, the duration of flights was reduced to compensate and achieve greater passenger throughput. A 'joyride' of just seven or eight minutes became the norm now.

This, then, was the scenario that brought one of the best-known flying companies of the twenties, Berkshire Aviation Co. – also trading as Berkshire Aviation Tours until the end of the decade – to south Lincolnshire in the late summer of 1923.

With origins similar to those of Summerfield's, Fred and John Holmes, from East Hanney near Wantage, Berkshire, left the RAF in 1919, bought a war-surplus Avro 504K for £450 and set out to create an air transport company that began joy riding operations in May 1919. Fred was a mechanic, apprenticed to A.V. Roe in 1912 and served with the RNAS in the First World War. Coincidentally, he had been an earlier aeronautical visitor to the region on the occasion of Freddie Raynham's flying exhibition at Hunstanton in July 1914 (see chapter 3). Fred was Raynham's mechanic for the Avro seaplane during that tour. John, generally referred to as Jack, was commissioned into the RFC as a pilot in 1916. Shot down in France while serving with No.19 Squadron early in 1917, Jack was repatriated when the war ended and after two years out of the cockpit, wanted to find his feet again in the aviation world.

They kindled such interest in flying in their locality, that a second pilot was sought. A certain self-assured young fellow by the name of Alan Cobham, also an ex-RAF pilot looking for a job, responded to their advert and joined the brothers as a partner, forming the Cobham & Holmes Aviation Co. Alan Cobham stayed with the company through its carefree, hectic first year. Always living financially from hand to mouth (an ordinary civil pilot in those days would consider himself fortunate if his salary reached £400 a year), the business expanded, perhaps a little too rapidly, to four aeroplanes (believed to be G-EAIB, EASF, EAHZ and EAKX) in 1920, extending its tour programme nation-wide. Another significant economic factor was that farmers, on whom they usually depended for suitable grass fields, began to cash in on this new income opportunity and pushed rents upwards. When financial backers let them down and cash flow waned, Cobham decided to part company and forge his own path.

Alan Cobham wanted, above all else, to make a career in the world of aviation. Having parted company with Berkshire Aviation in 1920, just such an opportunity arose for him to join the Aircraft Manufacturing Company (Airco) and this 'impecunious pilot' grasped the chance. Airco decided to expand its business into aerial photography, which was one area showing signs of growth alongside an otherwise contracting aircraft manufacturing industry.

The company converted one of the DH9 aeroplanes it manufactured to carry a photographer in the rear cockpit – minus its more usual warlike accoutrements – and it was to the newly-created and grandly named Airco Aerial Photography Department that Cobham reported as chief (and only) pilot. Aerial jobs kept the team very busy but a bombshell dropped when the aircraft building recession bit hard and Airco went into liquidation. This blow might have brought Cobham's intended career to a swift end but according to his biography, it was to be a fortunate turning point instead.

Geoffrey de Havilland, former chief designer at Airco, founded his own company and decided to hire a couple of DH9s and in turn hire them out, mainly to aerial photography companies and for air-taxi work. With this in mind de Havilland approached Aerofilms for work and the latter readily agreed but only on the clear understanding that Alan Cobham would do the flying.

De Havilland offered Cobham a job on the strength of this contract. The basis of the deal was that de Havilland would service and repair the aeroplanes and Cobham would charge him a flat price per flying hour and receive a retainer fee. Cobham accepted the deal and the first aerial photo sorties for the new company began in February 1921 using a DH9, G-EAYT. It was now that Cobham's skills as a pilot and astuteness as a businessman really came into their own. By living modestly on the ground and with clever, highly efficient flying in the air, coupled with meticulous sortie planning, Cobham was able to achieve extremely high productivity and profit in his flying operations. In his memoirs, Cobham considered 'this was the moment at which I achieved security in the world of aviation'.

It is against this background that we now see Alan Cobham making his first trip into fenland skies. The high quality results of his visit, as part of an aerial photography contract – with cameraman Joseph W. Edge in the back seat – can be seen within the pages of the *Boston Guardian* archives. Splendidly crisp views of Boston, Heckington, Sleaford, Wainfleet, Skegness and Lincoln, amongst others, appeared weekly between January and April 1923. Since G-EAYT was written off the British register on 2 October 1922, it seems reasonable to take the suggestion of 1921, in Cobham's memoirs, as the most accurate date for the working visit. Eight more event-laden years were to elapse before Alan Cobham returned to the region.

Also in 1921, having reduced its fleet to just two Avros, Berkshire Aviation Co. Ltd – as it had now become known – rode out its bad patch and in fact soldiered on until the end of the decade. It was not until the summer of 1923, however, that Berkshire Aviation itself reached out to embrace fenland audiences.

The composition of the tour staff at that time is not recorded but from press information, some possibilities can be deduced. J.D.V. (Jack) Holmes is known to have left the company during 1923 to join British Petroleum, while an advertisement in the *Spalding Guardian* of 22 September 1923 states the 'sole proprietor' of the tour to be 'Capt F J V Holmes, late of the RNAS'. As mentioned in chapter two, ranks and titles were often 'honorary' but it is reasonable to deduce this refers to the ex-RNAS Fred Holmes. It has been established that Fred was not a pilot and the use of his name in press reports seems to be purely in the context of the show in general. Capt. Holmes is reported quite clearly as the man who on Sunday 23 September 1923, as a finale after a two-week stay at Spalding, performed the hair-raising spectacle of walking on the wings of an Avro in flight. As a crowd-puller, this stunt had been a feature of the Holmes's show since its introduction by team member J.C.C. Taylor in 1920.

Local residents Robert Nicholson and Norman Massey provide colourful first-hand accounts of joy rides with the tour while it was in Spalding, and their accounts confirm the identity of one of the pilots. Recalling his trip, over half a century earlier, Mr Nicholson said:

As a youth, when aircraft were still a novelty, I flew with Mr Sparkes from a field in Clay Lake [Spalding]. The Avro carried two passengers in an open cockpit. I went up

Lt F.G.M. Sparkes of the Royal Flying Corps. (Mrs Irene Sparkes via Peter Green)

with a Mr Healey, whose son was landlord of the Robin Hood Inn on Bourne Road, Spalding. We had no goggles and in the fierce slipstream tears streamed down my face. There was a rainbow and it was a remarkable sight to view from above. Fields, houses, hedges and roads below us seemed painted across with those colours. Our pilot said he had been up 2,000 times and said this was only the second time he had seen such a sight. Wonderful!

Publicity photograph of Mr F.G.M. Sparkes of Berkshire Aviation Co. (Tony Wellband)

Spectators were admitted to the field for sixpence (2½p) and a 'quick flip round the circuit' cost 5s (25p) with a longer trip 'over and around the town' for 7s 6d (38p). Both gentlemen had their flights from a grass field in Spalding Drove, Clay Lake near Burr Lane railway crossing, and named F.G.M. Sparkes as their pilot.

Another ex-RAF lieutenant pilot, Mr Sparkes, had been in the barnstorming business since 1919 when, like Donald Sadler, he began his civil flying career with A.V. Roe. Working the Swansea district for Roe, he formed the Welsh Aviation Company in November 1920 when Roe ceased its own joyriding business earlier that same year. When times got tough after a year in business on his own, he wound up his company and joined Berkshire Aviation in 1922. During the 1920s he was a regular participant in air races held at various venues throughout the country.

Examination of press articles also reveals that two aeroplanes were engaged at Spalding. Photographic evidence, in the form of a souvenir card, shows one of these aeroplanes to be Avro 504K, G-EBFV. Analysis of individual histories of Avro 504Ks that found their way onto the British civil register shows G-EAKX, ex-H2600, to be the most likely second aeroplane at Spalding. As for the identity of the second pilot, analysis of the same source suggests A.L. Robinson is the most likely candidate. However, by the time the tour reached Wisbech (Levrington Common) on 1 August, Robinson's place had been taken by J.D. Parkinson and a third pilot joined the team for a short spell. The new man, Capt. O.P. Jones, was ostensibly 'on holiday' from his usual job, that of pilot with Instone Air Services, with whom he flew airliners regularly on the Croydon to Cologne route. Jones later became a much-respected airline captain with Imperial Airways and BOAC before his impressive flying career ended.

The most notable event in the hectic ten days at Wisbech came when no less than forty of Messrs Hickmans' farm employees were treated to a flight at their employer's

Berkshire Aviation Tours' team at Levrington Common, Wisbech, in August 1923. Seated from left are: Fred Holmes, his wife, and pilots J.D. Parkinson and F.G.M. Sparkes, with Avro 504s, G-EAKX and EBFV. (Bill Welbourne)

expense. Mr Hickman was allowed the trips at half-price in appreciation of his having halted steam ploughing of the stubble field on Roman Bank so that it could be used as a landing ground. These flights took place in a tea interval during the harvesting of other fields nearby, and the workers were conveyed to the landing ground by lorry. The *Wisbech Standard* reported, 'it was a novel sight to see a queue of land-girls in their distinctive bonnets waiting with a party of harvestmen to take a birds-eye view of their harvest fields.' The newspaper went on to relate the tale of a certain Mr Shaw who asked the pilot to loop-the-loop on his trip and then persuaded Flossie, a bonneted land girl who he had an eye for, to go up with him, which she did without the slightest hesitation, and said she enjoyed every minute of it! In all, the Berkshire team flew 400 men, women and children during that ten-day visit.

With its 1923 fenland sojourn beginning at March (the town) in July, followed by Boston and Wisbech in August, they moved on to Bourne and Spalding in September. Berkshire Aviation Tours then sent one aeroplane (EBFV) to Mr Winfrey's field in Sutton Bridge Road, Long Sutton, and the other to Bowser's Field in Hallgate, Holbeach. Here they remained from 24 September to 1 October; it was then the intention to move on to Crowland. There is some evidence of a programme in the King's Lynn/Heacham area around this time but details are still to come to light about that venue. Having proudly boasted of carrying 75,000 passengers without mishap since commencing business, this record was tarnished – fortunately without serious consequence – at Long Sutton on 24 September 1923.

Caught by a sudden gust of wind while taking off, EBFV, piloted by Mr Sparkes, staggered only momentarily but sufficiently for a wing to drop and scrape the topmost branches of an apple tree. Down it plunged, toppling the tree with its undercarriage as

Flossie and her workmates take a break from their labours at Levrington Common, Wisbech, in the summer of 1923. (Tony Wellband)

FLYING

The World-famed Berkshire Aviation Tours will give

PASSENGER FLIGHTS

FIVE SHILLINGS EACH

Including over and around the Town 7/6, from

SEPTEMBER 4th to SEPTEMBER 17th

at SLEAFORD ROAD, BOSTON

(By kind permission of H. Hall, Esq.)

On Sunday, Sept. 2nd, at 4 p.m. and 7 p.m. there will be a Daring Exhibition of ———

Stunt & Crazy Flying

FREE FLIGHTS by ballot and competition daily. ———

Two 110 LeRhone Avro Aeroplanes ——— will be in use all the time.———

ADMISSION 6d., CHILDREN 3d. (Including Tax).

YOU'LL BE UNDER THE EARTH ONE DAY; GET OFF IT ! ! !

'World Famous' Berkshire Aviation Tour brings spectacular aviation to Boston in September 1923. (*Boston Guardian*)

Souvenir postcard of Mr F.G.M. Sparkes, pilot of Avro 504K, G-EBFV, on tour in the Fens with Berkshire Aviation in 1923. (Robert Nicholson)

it fell. Luckily the tree also broke the fall and although the Avro was severely damaged, the pilot emerged unhurt and his two passengers, Mr and Mrs Ream of Bull Lane, Long Sutton, escaped with just a crop of painful bruises. Flights were suspended at Long Sutton pending the arrival of a replacement aeroplane and the remainder of the fenland tour seems to have passed off without further mishap.

In 1926 Berkshire Aviation Tours relocated their base from East Hanney to Witney in Oxfordshire and in 1929 amalgamated with Northern Air Lines to form Northern Air Transport Ltd.

What became of those personalities? In the Second World War Fred Holmes managed an aircraft components factory while Jack served in Fighter Command operations rooms with the rank of squadron leader. The former died in 1967 and the latter in 1980 after a lifetime at the forefront of aviation. Mr Sparkes became chief flying instructor (CFI) at London Aero Club in 1925. Emigrating to Canada in 1928, he formed his own air transport company in London, Ontario, later being appointed CFI at Border Cities Aero Club. Sadly he was killed in an air accident in 1934. A.L. Robinson left Berkshire Tours to join Imperial Airways in 1924.

Robert Nicholson eventually took over his father's bakery business in Pinchbeck Road, Spalding, and retired in the 1970s. For his next trip aloft he had to wait until his ninetieth birthday, when relatives arranged a trip in a Cessna 150 from nearby Fenland Aero Club. He thoroughly enjoyed the experience, enthusiastically reliving and comparing that flight with his first, almost seventy years earlier.

Skegness and Hunstanton, at opposite corners of The Wash, provided great opportunities for barnstormers to ply their trade to holiday audiences already in the mood for a good time. With this approach in mind, Capt. W.A. Rollason operated a de Havilland biplane (believed to be an Airco DH 6) from the Skegness showground, a scene of stirring displays by that intrepid aviator from another era, B.C. Hucks. One eager young passenger was Mr 'Deg' Smalley who recalled the day, 6 May 1923, when he went to the seaside by excursion train, from his home in Swineshead and 'blew' all his pocket money on a flight with Capt. Rollason. There is also a suggestion that a pilot by the name of Edward Milton also flew pleasure trips at Skegness during 1923, but it is not known if he was part of Rollason's team or operated separately.

'Walking the wings' was an essential ingredient of any 1920s and 1930s flying show. Just as ballooning had developed in the 1890s to embrace the risks of parachuting, so barnstorming added this sensation to its repertoire. Such stunts drew crowds, and crowds meant income for promoters. This hair-raising spectacle, though featured, was not however responsible for what the *King's Lynn News* headlined as 'Crazy Flying Tragedy At Lynn', in its 4 May 1926 issue.

Visiting the region on this occasion, for the purpose of pleasure trips and exhibition flights, was the Lloyd Commercial Aircraft Company owned by George M. Lloyd. Peterborough was scheduled as first venue on its intended fenland tour. Lloyd's one and only pilot, Capt. Arthur Orde Bigg-Wither, flew their one and only aeroplane, Sopwith Gnu, G-EAGP, north from Brooklands airfield. Bigg-Wither, a twenty-eight-year-old former First World War flyer, had fallen on hard times after the war, following jobs variously as a commercial traveller, and more recently, as a milliner in partnership with his fiancée. Despite a small private income he lived, according to a friend, 'in an extremely poor neighbourhood' (Waterloo Road, London). 'This flying job,' continued his friend, 'was Arthur's first really good opportunity in two years.'

Built in 1919 EAGP was one of only twelve Gnus built and this one had the distinction of winning the 400 mile Grosvenor Trophy race of 1923. Twenty-two-year-old stuntman George Lloyd had only recently acquired it from Southern Counties Aviation Co. of Brooklands. So recently, in fact, that the Air Ministry certificate had not been amended to reflect its new ownership. This situation may well have arisen because Lloyd needed a replacement for another pleasure-flight Gnu, G-EADB, also purchased from Southern Counties earlier that year that had stalled and crashed while landing at Horley, Surrey.

Handbills circulating in Peterborough announced that Capt. Bigg-Wither would give flying exhibitions and passenger carrying flights from a field in the Eastfield district of the city, between Sunday 25 April and Sunday 2 May. He actually arrived on Monday 26 April, landing in a field off Padholme Road, but no flights were made, as conditions were said to be 'unfavourable', presumably due to bad weather. Expressing his intention to return to Peterborough the aeroplane, curiously, was then dismantled and conveyed by train to King's Lynn. Here another flying exhibition was scheduled from Thursday 29 April to Sunday 2 May, but once again the weather played a part, delaying flying until the Saturday.

This company was indeed an operation 'flying by the seat of its pants' in the true traditions of barnstorming. Perhaps in need of income quickly and glad, no doubt, to be able to get airborne, Bigg-Wither, despite a strong wind, made a short publicity flight over

FLYING AT SPALDING.

THE BERKSHIRE AVIATION TOURS beg to announce that on *Sunday Next*, September 23rd, at **4-15 p.m.,**

Capt. F. J. V. HOLMES

will attempt the Sensational Feat of

WALKING THE WINGS

whilst the Aeroplane is travelling at 90 m.p.h.
You may have seen this American Film Stunt on the pictures.
Come and See It in Reality.
Free Flights for those who pay for admission to the field.

Passenger Flights from 5s.

Admission **6d.** (including Tax). Children Half-Price.

Walking the wings. Thrills in the air at Spalding in 1923. (*Lincolnshire Free Press*)

Daredevil Harry Willis walks the wings of Avro 504K, G-EBYW. (Cobham Plc archive)

Sopwith Gnu, G-EAGP, crashed at King's Lynn on 2 May 1926. (Peter Green)

the town that afternoon. Posters also proclaimed a 'Special Event' for Sunday afternoon. It was to be:

> an exhibition of crazy-flying and if conditions are favourable, one of the company will attempt the sensational feat of walking on the wings while the machine is travelling at 90mph.

Bright sunshine arrived with Sunday but a gale of a wind blew in with it. Undaunted, Arthur Bigg-Wither, watched by an expectant crowd of 2,000, began passenger trips at 2.30 p.m. After a dozen passengers had been taken for short, low-level trips in the three-seater aeroplane, a little difficulty was encountered in restarting the engine. Three quarters of an hour passed while Albert Hawes, their mechanic, attended to clean the plugs and Arthur chatted nonchalantly with his fiancée.

At 4.30 p.m. all was pronounced well. George Lloyd took up position in the rear cockpit, inviting a local man, Arthur Barrett, to join him for the next trip. Bigg-Wither took off in the blustery conditions and climbed high into the bright sky.

Soaring to 2,000ft over Gaywood Hall the pilot put the Sopwith into a series of steeply banked turns. Even at that height there were some among the watchers who were aghast at such a manoeuvre in so strong a wind. Excitement increased when the aeroplane steadied and Mr Lloyd was seen to climb out of the cockpit. Hesitantly, he made his way along the starboard wing, waving to the crowd below. A rousing cheer went up. Lloyd lay down on the wing, peered over the leading edge and waved again. Another cheer

Swinging the propeller of Sopwith Gnu, G-EAGP, before its last flight. (Peter Green)

rose from the ground. Now he inched his way back to the rear cockpit and to the relief of Barrett and the watchers, climbed in safely. Speaking later, Mr Barrett described what happened next:

> We continued to climb to do some stunts purely for exhibition purposes. The pilot and I talked to each other by hand signals. He made a looping movement with his hand and shook his head. I did the same, indicating we had agreed not to loop, for the wind was still fairly strong. He then made one or two steeply banked turns and signalled his intention to spin the aeroplane. I was a little apprehensive but did not mind and do not recall objecting.
> The pilot put the aeroplane into a right-hand spin. Just at the moment he appeared to try to flatten out, the aeroplane seemed to dive again – as if in an air pocket. When he did flatten out we were too low and it was obvious we should hit the trees.
> Seeing that crashing was inevitable I raised myself partly from the seat with my knees well up, hoping to be thrown clear. Mr. Lloyd, on the other hand, shouted: 'we are for it, old man!' and crouched low in the cockpit. It was all over in an instant. We hit a tree which broke our plunge and when the machine came to a standstill I heaved a piece of wreckage clear of my legs and clambered out.

Arthur Bigg-Wither died from the injuries he sustained. George Lloyd was seriously injured but lived to tell the tale. Arthur Barrett fell to earth and literally walked away with it, escaping with only facial cuts and bruises – a very lucky man indeed!

The aeroplane crashed into a corner of Hardwick Road cemetery grounds and it was in that same quiet corner that the final chapter of this tragedy was enacted. Being a Roman Catholic and having no close relatives in this country, Arthur Bigg-Wither was laid to rest in the Roman Catholic section of that cemetery, not twenty yards from the crash impact point. To this day, though, he is destined to lay in an unmarked plot, in the shadow of the very elm trees that contributed to his death.

With an emphasis thus far on mishaps, the reader may be forgiven for thinking that joy riding was indeed a hazardous business. In fairness though, it should be remembered that, in the context of this region at least, the total civilian deaths and injuries, while regrettable, was still only in single figures. By contrast, therefore, tens of thousands of passengers were carried aloft in this region – and elsewhere – without mishap during the twenty years between the wars.

In all that time in the region under discussion, only two passengers were killed and three injured, in addition to the deaths of two pilots. Penrose (British Aviation, volume 2) for example, records that in 1919 the large A.V. Roe Ltd team carried 30,000 passengers in four months. By 1923 Fred Holmes claimed, according to the *Spalding Guardian*, that his company had flown 75,000 passengers; during his stay in the town he was certainly averaging twenty flights per day. Tens of thousands of passengers were joyriding every year.

The very nature of barnstorming made it a commercially tenuous way of life, being at the mercy of the weather, with expensive equipment to operate during a time of economic hardship. Nevertheless it was undoubtedly instrumental throughout the land in enhancing public awareness of air transport and enriching people's lives through the exhilaration to be found in watching and participating in flying. Examination of local newspapers for the ten years up to 1929 indicates that joyriding provoked a progressively declining interest to newspaper editors, unless a major celebrity or mishap was involved.

Editorial and public interest were both rekindled, though, by the arrival on the aeronautical scene of the 'Record Breakers', and to some extent a parallel can be found here with the exploits of Charles Green and his contemporaries during the 1830s.

Foremost among the procession of men and women who became famous for pitting their wits against the aeronautical unknown, stands Sir Alan Cobham. This, by no means, detracts from the achievements of others such as C.D. Barnard, C.W.A. Scott, Amy Johnson, Bert Hinkler, Jim Mollison et al, for their skill and courage, too, is unchallenged. In Cobham's case, though, he made his name comparatively early on, then capitalised on his other great asset, his personal marketing flair, and his star remained in the ascendant for a much greater time.

A common factor running through conversations with men and women who flew with or watched barnstormers and air shows between the wars is that almost all say they 'went up with, or watched, Cobham'. Subsequent analysis of their stories shows that many actually did not do so with him but with one or other of his competitors. To these people though, any air show meant 'Cobham', such was this man's renown

A cheery wave from a knight of the air, Sir Alan Cobham. (Colin Cruddas)

It will be remembered that Cobham parted company with his Berkshire Aviation colleagues well before their 1923 visit to fenland. The great man made a brief foray into the Boston area during 1920, but nearly a decade passed before he returned in 1929, this time not as an impecunious pilot struggling to make a living but as a knight, riding a 'big blue charger'. Fresh from a string of record and route proving flights, in 1928 he sought another outlet for his energies.

His new crusade was to create as much enthusiasm as possible, throughout the land, for the total concept of civil aviation. It has already been established that every year so far throughout the decade, many tens of thousands of ordinary people were exposed to aviation by the barnstorming fraternity. It must be said, therefore, that Cobham's strategy was mounted on a much wider front and directed at public bodies in addition to, rather than solely at, private individuals, and he wished to promote aviation specifically among the youth of the country. As he put it in his own words, 'I wanted to make Great Britain air-minded.'

With this aim in mind he conceived two schemes. First of these was his 'Municipal Aerodrome Campaign' of 1929, followed by the 'National Aviation Day Campaign' from 1932 to 1935.

DH61 Giant Moth, G-AAEV, *Youth of Britain*, flown by Sir Alan Cobham on his Municipal Airports Tour of 1929. (Cobham plc archive)

In common with hundreds of local authorities, Spalding Urban District Council was the recipient of an application from Cobham, inviting the town's participation in the Municipal Aerodrome Campaign.

Spalding Council received a letter during April 1929, in which Sir Alan explained he had originally approached 'practically every authority throughout the country on the subject of establishing municipal aerodromes, since July 1927'. It is evident from this letter that his implicit objective was to enhance his own Cobham Aviation Consultancy business – but who could blame him for that? It is fair, though, to credit him also with a sincerity of national purpose, as he went on to outline publicly and with fervour his opinions on the necessity to develop air transport in Britain.

Financial backing and a free supply of oil for the planned twenty-one-week duration tour was provided by Sir Charles Wakefield, chairman of British Petroleum (BP). Cobham was thus able to purchase an aeroplane appropriate to his purpose, a de Havilland DH61 Giant Moth, G-EEAV, one of ten examples built and one of only two operating in Britain. With a 52ft wingspan, the Giant Moth was a large (7,000lb), single-engine biplane with an enclosed passenger cabin immediately behind the engine and a loftily positioned open cockpit aft of the cabin. Power plant for this ten-seater airliner (in Cobham's configuration) was a 500hp geared Armstrong Siddeley Jaguar VIc engine, giving a top speed of 132mph and a landing speed of 46mph.

Sir Alan specified that the field he required must have a minimum landing run of 400 yards in all directions and would therefore encompass an area of some thirty-three acres. Furthermore, it needed a level surface and no obstructions on its boundaries. At least in the fens of eastern England he stood a good chance of finding such venues! Cobham employed former First World War pilot Dallas Eskell as his ground manager, whose small team, including J.A. Dick and a Mr de Menthe, surveyed potential landing sites all over the country, negotiating with landowners and councils for the use of those deemed to fit the bill and briefing them prior to each event.

Following Spalding Urban District Council's enthusiastic response to his proposition, Sir Alan's representative, on 19 April, inspected a field in Giddon's Lane, Wykeham, three miles north-east of the town, kindly offered by local farmer, Mr R.T. Proctor. Although cropped with clover, it was pointed out that the field would be cut and suitable for landing during the next three weeks but thereafter would not be fit again until July. It was in his letter of confirmation to the UDC that Cobham announced his intention to honour the town by beginning his nationwide tour in Spalding on Friday 17 May 1929 – although it was, in the event, actually preceded by Oxford and Bedford.

At Wykeham the momentous day dawned clear and bright. In a corner of one of Proctor's fields, farm labourer George Hare sat chewing clover. Chewing was a habit of George's but today he did it without paying much mind to it, because his attention was focussed on trying to work out why so many people were visiting that particular clover field that morning. Some came in cars, some on motorcycles, others on pushbikes and even more on foot. Personally, George could see little of interest to warrant such an exodus.

Of course there was that balloon-looking thing on a post over in the other corner of the field. It looked, to him, just like an inflated sausage skin. Squinting against the light he could also make out a rope running on short posts along another side of the field and – oh yes – there was a BP petrol lorry too. Everyone seemed to be waiting for something and craning their necks expectantly to the sky.

Mr Hare could not see very well from where he sat, so he got up and trotted through the clover to the other end of the field. As he jogged along he saw people laughing at him but quite why, he could not imagine. Then he noticed the crowd was pointing behind him so he stopped and turned round. Out of the distant sky came what he at first took to be a large bird. Straight as an arrow it drew nearer and with it came a droning noise that grew louder and louder. Standing there, it dawned on him that the 'bird' was one of those new-fangled aeroplanes and bird or not, he would not like it to sit on him! Thinking thus, George turned on his heels and did what he described as 'a bunk!'

Making two circuits of the landing field, indeed like some giant bird, Sir Alan brought the blue airliner in for a graceful touchdown and taxied right up to the crowd. He jumped down from the cockpit to be greeted heartily by the Urban District Council chairman, Councillor R.S. Donington before being introduced by the clerk, Mr Raymond W. Hastings to other members. His short speech of introduction repeated the objectives of his tour and after many pleasantries about the hospitable welcome he set to work on the business of the day – to fly.

Sir Alan Cobham, centre, with manager Dallas Eskell, third left, and members of Spalding Urban District Council prior to him taking them aloft in the DH61 Giant Moth – a ritual he performed at venues all over Britain in 1929. (*Spalding Guardian*)

'Splendid! I can recommend it!' was the verdict of Councillor Donington upon his return to terra firma. As chairman of the council it was his privilege, together with his colleagues, to be taken up on the first (free) flight of the day. Thereafter the general public queued to take their turn, for the sum of 10s (50p) per head, in this pale blue and white aeroplane that carried the name 'Youth of Britain' on its nose.

It was Cobham's declared aim to fly as many schoolchildren, free of charge, as possible at each venue on his tour. It is not known what selection criteria were used but for every lucky child there must have been many disappointed ones. Among the latter was Peter Sanderson who lived on Roman Bank on the route from town to the airfield. Being, from a small boy, an avid aeroplane enthusiast he recalled that day with sadness, for he had contracted measles and was confined to bed. It was a bitter blow, made all the more hard to bear because he could hear the crowds passing by his house and the sound of an aeroplane droning to and fro overhead.

Sir Alan was kept busy all afternoon, taking up a total of eighty schoolchildren and ninety other paying customers by the time the arrival of dusk put an end to flying operations. The *Spalding Guardian* summed up this momentous day thus:

The visit of Sir Alan made a deep impression on all those who had the honour of being piloted by him. It is to be hoped that, when the whole of England is linked up by aerodromes, Spalding will be one of the towns with a landing ground.

The next day Sir Alan flew to Boston. It was customary for his advance representative (Mr J.A. Dick, in this instance) to brief the local authorities about each visit. Although council members had gone to great pains to ensure arrangements for their own free flights were all in hand, confusion reigned about the free flights for local schoolchildren – a key element of the programme. Boston's Education Committee chairman claimed to have no knowledge of any offer of free flight places for schoolchildren. This unfortunate situation was brought out into the open when the tour manager arrived in town early that Saturday morning (18 May).

That Boston schoolchildren might be deprived of this wonderful opportunity naturally caused a furore and word was hastily sent out to schools – which were closed – so that invitations could be delivered by hand to the lucky youngsters before Sir Alan arrived. It has been suggested that tickets for the children's flights – there was usually a quota of about seventy free tickets per town at first but this was later increased to ninety – were awarded as prizes for essays entitled 'Flying In Relation To The Empire'. Local education authorities would therefore be highly involved with organising this element and the idea was that they, together with administrators in the Wakefield organisation, would judge and select the winners. That was quite some task, particularly as it would all have to be done and dusted before Sir Alan reached each venue! Whether this process was fully implemented seems to be in some doubt and in view of the logistics involved, one can understand why it might have fallen by the wayside for a more expedient method of selection – a simple reward for current academic ability perhaps – at local level. Events in Boston seem to support this latter view.

Choosing which lucky pupils should have the honour of flying with Sir Alan had to be dealt with swiftly and Boston High School for Girls, for example, delegated the job of drawing up a list of girls at random, to its head girl. Sixteen-year-old Laura Nutman (née Hibbins) was one name on that list. She recalled that she was completely unaware of the impending visit.

> I don't remember reading anything about the visit in the newspapers, but I suppose those sort of things weren't of much interest to a young girl in those days. I jumped at the chance though when it came. There was a knock on the door of my home in Cheyney Street and the invitation was delivered with instructions to go to the flying field straight away.
>
> I cycled the two and a half miles to where the aeroplane was parked on Walter Day's farm on Sleaford Road. Alan Cobham and his manager were there and we went up in groups of ten at a time. Of the take-off and landing I remember little but the aeroplane in flight was noisy and rattled a lot. The flight lasted about fifteen minutes and seemed so short at the time but we all thought it was wonderful.

Each group had its photograph taken alongside the aircraft before going aloft and Laura still treasures that souvenir of her trip with Cobham.

An item from the minutes of Peterborough City Council dated 22 February 1929 gives further insight to the preliminary planning of his schedule for that first campaign, which was by no means all plain sailing:

Visit of Sir Alan Cobham to Boston, 18 May 1929, during his Municipal Airport Tour. Tour manager Dallas Eskell helps pupils of Boston High School for Girls to board the Giant Moth. From left to right: Jenny Ablard, Rose Boardman, Rose Sykes, Betty Coy, Doris Burton, Olive Waddington, Ethel Coy, Mabel Thompson, Dorothy Bentley and Laura Hibbins. (Martin Nutman)

> Sir Alan Cobham applied to the council for permission to land in a field in Peterborough as part of his 1929 Tour. The City Chief Engineer is to reply to indicate the best field for this purpose. The Mayor undertook to give Mr Cobham all assistance.

But later that year the *Peterborough Standard* of 13 September, reported, 'Sir Alan Cobham, who was to have visited Peterborough on Friday October 4, has had to cancel his visit owing to the difficulty of obtaining a suitable landing ground.'

At Westwood, on the edge of the city, was an RAF Aircraft Reception Park which Cobham thought appropriate for his purpose and made application to use it. Unfortunately for the city the Air Ministry had other ideas. Cobham was obliged to write to the mayor informing him that the RAF had told him it was not convenient to use their Reception Park as 'it had been seeded down'. Sir Alan regretted that unless an alternative site could be found he would have to cancel his visit. Although two other fields were examined, neither proved satisfactory and the visit was indeed called off. Peterborough's loss, it seems, was Spalding's gain.

Although, shortly after the visit, Spalding UDC set up an airways sub-committee to examine the question of an airport, its implications and Cobham's involvement, nothing came to fruition as a direct result of his visit.

At the end of his gruelling 1929 tour the statistics speak for themselves. In twenty-one weeks Sir Alan and his ground team of about six people, visited 110 towns and cities, made about

No time to lose! With the engine running between passenger loads, Sir Alan still finds time to deal with eager reporters. (Cobham plc archive)

5,000 flights during which almost 10,000 schoolchildren, 3,500 local government dignitaries and 36,000 paying passengers were taken aloft. For complex business reasons G-AAEV was sold to Imperial Airways and Cobham even delivered it to Rhodesia for them where, sadly for so special an aircraft, it was wrecked by an Airways pilot in a crash in January 1930.

As for achieving the campaign's objective, there was no grand municipal airport in Spalding but it will be seen in a subsequent chapter that a small commercial flying company operated from the Spalding Marsh locality, not so very far from the field used by Cobham, for a few years just after the Second World War.

Meanwhile, back on the barnstorming circuit Boston remained as popular a port of call for barnstormers in the 1920s as it had been throughout the nineteenth century. These small aero companies came and went with monotonous regularity and few of those that 'took off' at the start of the decade were still flying at its close.

Captains Edward Jordan and L.W. Hall were ex-First World War pilots formerly with the famous Cornwall Aviation Co., but when it felt the economic squeeze their services were no longer required, so they set up in business as Western Aviation Ltd and continued to earn a living doing what they did best – flying for the thrill and pleasure of it.

After a hectic summer season at Cleethorpes, September 1929 saw the pair arrive in Boston with two Avro 504K aeroplanes, G-EBQR and G-EBXV. Sherwood Avenue, off the Sleaford Road, was the site of their temporary airfield and both aeroplanes were kept busy from 10.30 a.m. until dusk throughout the weeklong stay, flying passengers around Boston in the traditional way. Among the many 'first-timers' one enthusiastic Bostonian recalled:

It was a glorious afternoon when I made my way to the airfield and I anticipated my forthcoming adventure with great relish. Capt Jordan took me out to the aeroplane which was constructed to carry two passengers and I chose to sit in the one at the back of the rear cockpit to obtain what I thought would be a better view. I regretted my choice a little later as, being less protected by the windshield, the wind whipped across my face, making it sting with its force.

The engine started and the machine taxied across the grass gaining speed until, before I knew it, we were up in the air. Peering over the side I watched the ground drop away below and gradually the shadow of our aeroplane got smaller until it finally disappeared. Circling Boston Stump I could see the brown water of the river Witham enclosed by housetops of all colours from red to dull grey and all those framed by the green of the countryside.

There was no feeling of discomfort; the wind on my face set every nerve tingling but it was cool and fresh and were it not for the noise of the engine and the wish to take in all of the lovely view below, the gently moving aeroplane might even have made me sleepy.

Suddenly the engine noise stopped, the nose of the machine dipped towards the ground and it seemed as if the earth had been propped up at an angle. Now the engine roared back into life. We climbed once more, steeper and steeper until now the land seemed almost to be at the tail end and nothing in front but a mixture of canvas wing and empty sky [a spin coming!]. A swoop downwards and there was the ground again, rocking slightly through the arc of the propeller. Down, down we fell, with the buildings and fields spinning madly bringing a curious tingling sensation in my stomach! Flattening out, the pilot flew towards Bargate Bridge, banked the biplane steeply onto one wingtip, circled the town again then headed back to the landing ground. We landed with scarcely a bump and taxied back to the temporary office. It was all a delightful experience, long to be remembered and I was very sorry it had to come to a close.

There was to be a programme of 'stunt' flying by the two airmen during the final Sunday, before they departed for Horncastle where, if the usual Air Ministry show licence was forthcoming, they would offer their delights to a new audience.

'There will be no flying during hours of divine service on Sunday morning.' So promised the advertisement for passenger flights printed in the *King's Lynn News* of 8 October 1929, on behalf of Surrey Flying Services.

Founded at Croydon in 1922 and one of those few remaining in business by the end of the decade, Surrey Flying Services variously operated a dozen smart blue and silver Avro 504Ks and its five-seater variant, the Avro 536, throughout the company's life. One of the former, believed to be the 130hp Clerget-powered G-EBDP, was despatched for a three-week sojourn in King's Lynn. From Wednesday 9 October to Monday 31 October, passenger flights, stunt flying and walking-the-wings were the tempting aerial treats in store from a temporary airfield at North Lynn farm, Estuary Road. The pilot for the show was yet another First World War veteran, Fg Off. Earl B. Fielden, a Yorkshireman – not to be confused with the Prince of Wales's (Edward VIII) personal pilot Flt Lt Edward H. (Mouse) Fielden.

Avro 504K, G-EBDP of Surrey Flying Services visits King's Lynn in October 1929. (Ray Wilson)

Joyriding companies seem to have been starting up and closing down with increasing regularity as the decade wore on. This accounts for pilot names becoming associated with several different aviation companies during their flying careers – a somewhat confusing situation for the erstwhile researcher. Capt. A.M. West, for example, is noted as visiting Downham Market in September 1930 with the Wolverhampton Aviation Company. This gentleman joined Cornwall Aviation Company in 1931 and is also seen photographed in front of Avros of the North British Aviation Co during its 1931 Lakeland Tour.

Under the banner of Wolverhampton Aviation, West gave passenger flights and aerobatic displays from a field at Broom Hill near Downham Market (now built over as Fairview Estate) in late September 1930, flying Avro 504K, G-EBKR (ex-RAF E3382). Local resident Mr G.D. Hewitt remembers the day he and his pal went to the field, hoping to have a flight. It was 29 September and they were told the aeroplane had just undergone a minor repair and awaited an air test. Seeing disappointment on their youthful faces, West said they could go up with him on the air test but at their own risk. Rather than put them off, his offer added the prospect of more excitement so, paying 2s 6d (12½p) each – half the usual price – they went for a longer than normal trip over the countryside around Downham.

The ubiquitous Avro 504 was not the only type on the barnstorming scene in those days. North-Eastern Aviation Co., for example, operated a Spartan Three Seater that, over the Easter bank holiday of 1932, was brought to the same field in Estuary Road, King's Lynn. The chief pilot during the four-day stay was our old friend, Capt. L.W. Hall who, it will be recalled, was last heard of in Boston in 1929 with Western Aviation. Having moved on yet again he was reassuring his customers that he had been flying since 1915 and had flown 20,000 passengers, 'without mishap' since 1928. What his pre-1928 record

Souvenir postcard of Avro 504, G-EBKR, and its pilot Mr A.M. West of Wolverhampton Aviation Co. at Downham Market. (Ray Wilson)

was like was not forthcoming! In addition to the usual passenger flights for 5s (25p), the programme included 'stunt' flying every day.

From a slightly bizarre story carried by the *Lynn News* more than a year later and the description of Capt. Hall as 'chief' pilot, it may deduced that he shared piloting duties with one Mr John Rogers. The latter's name came to light when smuggling by aeroplane was alleged in a case – the first of its kind – at London's Mansion House police court in August 1933.

Two men, one of whom was a pilot named as John Rogers, appeared on summonses alleging attempted fraudulent evasion of customs duty on saccharine (a sugar substitute). Described as 'a well-known pilot' the *Lynn News* report commented that he had 'thrilled many Lynn people last year with a remarkable display of crazy-flying at the North Lynn field'. The prosecution alleged 63lb (28kg) of the sweetener saccharine, contained in twenty-three tins, was flown illegally from Rotterdam to England by Rogers and passed on to his friend. In his defence, Rogers claimed he became involved in the smuggling by a spirit of adventure and that he was very hard up at the time and hoped to make a profit. Defending counsel said Mr Rogers 'was a brilliant pilot with some five or six pounds a week and his friend had lost his capital through the confiscation of his aeroplane, bought for hire-work'. Each defendant was fined £100 with 10 guineas costs in respect of the non-declaration of imported goods.

All this was in the future, however. In March 1932, at least, those barnstormers left a lasting impression upon a small boy who was taken, by his parents, to see Capt. Hall and his aeroplane. Interviewed by the *Lynn News* later he gave an excited twelve-year-old's description of his first flight:

> I did not think I should go up in Capt Hall's aeroplane until just a few minutes before I went. Mother said she was sure she would not let me go up for it was not worth all that money. But Father and me were watching them flying and he said: 'Would you really like to go up?' and I replied 'Yes, ever so'. Before I hardly knew where I was, we were sitting in the aeroplane, him in front and me at the back [the Spartan had two tandem seats in the front, passenger, cockpit]. I did not mind this because the wind was very strong and he kept it off me so I could see out better.
>
> They swung the propeller, like you see them do it on the pictures [films] and it made such a noise going round that we could not hear ourselves talking. The aeroplane moved forward over the field and it was ever so funny because the driver was behind us, not like in a motor car, so we could not see him. It bumped a bit, went faster and then got smooth. Father shouted 'We are off the ground' and everything seemed to go away from us instead of the other way round as you would expect. We did not seem to be going fast now but the wind from the propeller was strong and that showed we were going at a good rate. It was a job to look out because of the wind so I got close behind my father and squinted over the side. Everything looked tiny and neat, with buildings like doll's houses and fields like patterns on a counterpane. Railway lines seemed ever so tiny and the dock was like a swimming bath, all green.
>
> We started to come down suddenly and I could just imagine going head first into the river and I was glad father was in front. The aeroplane started to turn and slope terribly and I hung on tight to the side. It didn't seem to want much more before I might fall out. I was not frightened really but it gives you a funny feeling inside, like a switchback ride. The field was getting nearer now. As we came down I thought we should never clear the bank and hedge but we did and hit the ground quite hard. It has good springs though and it's no worse than some cars or buses on a bumpy road.

The sheer delight of this child's experience can be sensed through his story but one cannot help guessing that strong words may well have passed between mother and father upon the latter's return to earth!

Now the scene moves to Littleport, later in that summer of 1932. Just a year earlier, a flying club was formed at Skegness, under the auspices of Eastern Air Transport Company, of which more later. Like all such companies of that era, its aeroplanes were employed at all manner of flying events. One such opportunity arose when the Skegness-based company was asked to provide aerial entertainment for the Littleport (Isle of Ely) Horticultural and Foal show. It requires but a small effort to imagine being back in the era of the 1890s country shows described in chapter two.

Flt Lt Spencer RAF, a flying instructor at RAF Cranwell, was engaged by Eastern Air Transport to pilot a DH60G Gipsy Moth, G-AAKM, at the show, selling passenger trips and acting as carrier for a parachutist. Disaster marred the display that day, 26 July 1932. Parachutist Robert Hopkins, just twenty-two years of age, clambered into the front cockpit of the Moth and Spencer took off. Climbing to 500ft, circling the showground, Hopkins was seen to stand up then plunge headfirst between the main- and tailplane. A cry of horror rose from the crowd. As Hopkins fell clear of the cockpit his parachute streamed out and became tangled with the starboard elevator. His trapped body could be seen suspended behind the Moth, the balance of which was now very seriously upset. It plunged earthwards, bucking violently as Spencer wrestled with the controls. Succeeding in arresting its headlong dive, the Moth finally stalled in from 100ft, crashing into a cornfield near the showground. Hopkins met a terrible death and Spencer, badly injured but alive, was cut free from the wreckage. Rushed to hospital 'with terrible injuries and in critical condition' he was, by the following week, reported to be recovering well.

Returning a verdict of 'accidental death', an inquest jury did not dwell on the precise cause of the accident but it seems possible that the parachute ripcord was pulled just too quickly. A certain 'one-armed Sqn Ldr Spencer' was noted in local newspapers in 1937, as being a pilot and a regular visitor by Tiger Moth to a relative in Heacham. Could this be the same man bearing the scars of his 1932 accident?

On a more pleasant note, this barnstorming era can be drawn to a fitting close by demonstrating the role of women flyers who, despite a great deal of prejudice against women in aviation in those days, took a lead from Amy Johnson and carved out their own niche. The East Midland region can be proud to have benefited directly from the exploits of two such women who went on to achieve even higher, national, esteem.

One of two daughters of a Tory MP, Pauline Gower learned to fly in 1930. While taking lessons at London Aero Club (for £1 an hour in those days) she befriended Dorothy Spicer, an equally determined young lady aspiring to enter the male-dominated trade of aircraft engineer. So well did they hit it off that before long this pair resolved to forge a career together in aviation. Pauline took the role of pilot and Dorothy became the ground engineer of this budding partnership.

Littleport (Cambridgeshire) Annual Sports Day was marred by the tragic death of parachutist Robert Hopkins and the crash of DH60 Moth, G-AAKM, on 29 July 1932. (Ray Wilson)

By the late summer of 1931, both had achieved their basic qualifications, which they soon expanded to commercial standard in their chosen field. During the following year they jointly purchased a light aeroplane and set up a joyriding business near Pauline's home close to Sevenoaks in Kent.

Aviation was a hard and, generally speaking, male-dominated world for a woman to pursue a career in, and not only because it was an expensive occupation. By way of illustration, a RAeC official at the time observed, 'Women make good pilots but I would not say that they are as capable as men in an emergency, because they do not seem to have that self-command which men possess.'

It was twenty-one years since the first pilot certificate had been awarded to Lt. Col J.T.C. Moore-Brabazon when, in August 1931, the 10,000th Royal Aero Club aviator's certificate (first mentioned back in chapter three) was issued – to a woman! Membership of light aeroplane clubs in Britain that year stood at about 8,000, a figure that had doubled from the previous year. The RAeC pilot certificate had to be obtained by satisfactory completion of flying tests before a pilot could be issued with an Air Ministry 'A' licence, which then entitled a man or woman to fly an aircraft for private purposes. Of the 954 'A' licences issued in 1930, 887 went to men and 67 to women. If a pilot wished to fly with paying passengers or to fly freight as a business then it was also necessary to obtain an Air Ministry 'B' licence, and between 1926 and 1930 only 330 of these had been issued. In 1930 itself, of the 110 'B' licences issued, 109 went to men and just one to a woman – Winifred Spooner.

Joyriding was a hard school to grow up in but Gower and Spicer persevered. In the next few years, as their experience grew, while Air Trips Ltd provided a basic source of income it was supplemented by appearances on the air pageant 'circuit' that was now gaining momentum. After trying their hand with Modern Airways Ltd – which traded as the 'Crimson Fleet' – in 1932, they were invited to join Henry Barker & James King's Air Pageants company for its 1933 season. This organisation operated some of its air shows under the banner of British Hospitals Air Pageant, fronted by C.W.A. Scott and it was with this team that Pauline and Dorothy visited fenland skies for the first time. They formed part of the passenger flying team for Scott's show and went on to repeat that role with Barker & King's British Empire Air Display of 1936.

Tiring a little of the nomadic barnstorming life, these 'women with wings' spent two idyllic summers, 1934 and 1935, operating their own air taxi and pleasure flight business in the seaside town of Hunstanton. Renting a field from Mr Searle, near what is now Searle's Caravan & Holiday Park in the South Beach area of the town, they lived a frugal existence in an old caravan, with their two dogs; a Spartan Three Seater I aeroplane, G-ABKK, grandly named *Helen of Troy*, was their sole asset. Mr D.W. Herring, nineteen years old in 1934, remembered their first season at Hunstanton.

> I went to the field just after they arrived. When Pauline landed she asked me if I wanted a trip. It was to be a longer flight than the usual joyride as she had been engaged to fly over the town towing an advertising banner. Second passenger on that trip was to be the crossing keeper from the railway line that ran alongside the field. Naturally we jumped at the chance of a free trip.

For a £1 fare, Air Trips Ltd offered a return flight across The Wash to Skegness and back, while joy rides around Hunstanton cost 5*s* (25p). An added thrill could be had by 'bombing' a speedboat, owned and usually driven by Mr Searle, with bags of flour.

Later, these enterprising ladies engaged the services of Bruce Williams as parachutist and general handyman to the team. Special events, to attract holidaymakers, were staged usually on Sunday afternoons and included 'crazy' flying, the use of wireless to control airborne operations, parachute drops and the inevitable flour-bombing routine – this time with Bruce Williams on the receiving end in a car. As many as 200 passengers would be taken up on a good day and by this date the women claimed to have flown nearly 18,000 people since they went into business.

Returning to Hunstanton Pauline and Dorothy acquired a new caravan and an additional aeroplane, a DH83 Fox Moth, G-ADNF, for their 1935 season. Business was brisk and special displays, sometimes mounted in mid-week, often attracted more than 1,000 spectators to their little airfield.

The season seemed to be going pretty well until one Sunday in August, when the *Lynn News* reported in a most dramatic fashion, 'By a wonderful feat of skilled piloting, Miss Pauline Gower saved two passengers and herself from death.'

It was on the afternoon of Sunday 11 August, when Pauline tucked her Fox Moth into the north-east corner of the field to give her the longest, 300-yard take-off run. Opening the throttle the aeroplane gathered momentum but its engine appeared to be losing revs. Faced with the prospect of crashing into an earth bank, Pauline took decisive action, hauled the nose up, staggered over the bank and dropped into a deep swamp on the other side. First on the scene was a *Lynn News* reporter, there for a routine story but now facing the prospect of an even better one. He found Pauline and her two male passengers, Roy Ball and another, unidentified, man lying on the grass on the far side of the swamp, from which they had crawled, shaken but with no more than a few scratches between them. Both men had escaped through a tear in the side of the cabin left by the departing engine, while Pauline had scrambled clear of the open cockpit at the rear of the passenger compartment. Looking somewhat worse for wear, the poor Fox Moth's nose was buried, it suffered a bent propeller, the undercarriage was crushed and submerged and the cabin almost completely destroyed.

Undeterred and in time-honoured fashion, plucky Pauline Gower was back in the air only an hour or so after her crash, first taking up Bruce Williams for two parachute drops then joyriding passengers in the Spartan Three Seater until dusk fell that day. The ever-practical Dorothy Spicer soon had recovery of the Fox Moth in hand, first removing the engine and undercarriage then using its folding wing facility to ease removal from the marshy ground. This latter task was achieved with the aid of a breakdown crane that made light work of dragging out the stripped down aeroplane.

Pauline spent that winter studying for her navigator, R/T and instrument licences, which she duly acquired, while Dorothy by now had qualified to a sufficiently high level to undertake both airframe and engine overhauls for their aeroplanes.

In 1936 they returned to the air pageant circuit, flying with Tom Campbell Black's show (of which more later) but the weather was miserable that summer and having had their fill of flying for the time being, both women decided to quit when that season ended. It

Two aspiring pilots admire the Spartan Three Seater, G-ABKK, owned and flown by Pauline Gower at Hunstanton, 1934. (Ray Wilson)

was during this period that their 'old faithful' ABKK was written off in a take-off collision with Avro 504 G-AECR, at the Campbell Black show at Westwood Heath in Coventry on 10 May, but they acquired a newer three-seater II, G-ACAD, with which to finish the season. Spartan Three Seaters were one of the mainstays of the pleasure flight business throughout the 1930s, and as late as 1939 examples such as G-ABET and ABKJ can be found in that role across the Wash in Skegness where they were operated by Kennings Ltd and also G-ABKT with Peterborough Flying Club at Horsey Toll airfield.

Dorothy Spicer ran an 'aerial garage' for a while but her heart was not in it and she sold up after a year. When war came Dorothy found an outlet for her engineering skill at Farnborough. Tragically, she was killed with her husband in an air crash in South America shortly after the war ended. Pauline Gower's career in aviation reached its zenith with her contribution to the war effort through untiring, energetic and inspiring leadership of the women's section of the Air Transport Auxiliary (ATA). Her role and that of the other women pilots of the ATA, is recorded in the book *The Forgotten Pilots* by Lettice Curtice. It is coincidentally tragic that Pauline Gower, too, died at an early age, in 1947 while giving birth to twins.

These two women will always remain shining examples of achievement in the annals of British aviation history and their story brings to a fitting close this barnstorming era, the likes of which would not be seen again.

CHAPTER FIVE

Pageants in the Air

Captain Charles Douglas Barnard was a record-breaking aviator whose name ranks alongside those of Cobham, Hinkler and Kingsford-Smith when such men were pushing back barriers to air travel in the late 1920s. In that heyday of long distance record flights, having carved out a reputation for himself, Charles Barnard capitalised on his success and the surge of interest in aviation by creating, in 1930, an air travel company called C.D. Barnard Air Tours Ltd. This was an organisation set up to tour major towns and cities in Britain and present air shows to the public.

In some respects Barnard emulated Sir Alan Cobham's 1929 tour, while in other respects he upstaged Cobham. Heavily sponsored by Sir Charles (later Lord) Wakefield, the Castrol oil magnate, Cobham's 1929 tour was based around a single aeroplane. While it can rightly be argued that Cobham's tour had much more far-reaching objectives than his barnstorming predecessors it is still more appropriate to see it as heralding the close of the barnstorming era. Charles Barnard, on the other hand, created what might be seen as the first of the multi-aeroplane, multi-act, one-day 'Aerial Pageants' that were to characterise the aviation scene throughout the 1930s.

By later standards, however, it was a modest team that embarked on what Barnard proclaimed as 'The World's First Air Circus', centred around that famous 'old girl' among record breakers, the ten-seater Fokker FVIIa G-EBTS, formerly named *Princess Xenia*. Now re-engined and renamed *The Spider*, it was the aeroplane in which he, the Duchess of Bedford and Robert Little set many records a year or so earlier. In support was a Cierva C-19 Mk III autogyro, G-AAYP; a Spartan Three Seater I biplane, G-ABJS, and a Hermes-engined Avro Sports Avian, G-AAXH. In addition to himself as chief pilot, the services of Reginald Brie, an autogyro expert; Capt. Edward Ayre, Mr Len Stace and T. Neville Stack were engaged to fly the biplanes, while John Tranum provided the daredevil ingredients of wing walking and parachuting. During the tour, the above men and machines were augmented or replaced from time to time by other pilots and aircraft.

It was on 2 April 1931 that the 'circus' set off on a projected tour of 150 venues, arriving in the full glare of publicity in Spalding on 8 May. This, the first of its fenland venues, was due to the co-operation of Mr R.T. Proctor of Wykeham. His vast holding of large, flat, arable fields around his Wykeham Abbey home, a couple of miles from town, offered some of the best landing grounds in the region. Cobham flew from here in 1929, as did all subsequent Spalding air shows in the 1930s.

Being a Friday, a working day, attendance in the morning was small. Nevertheless, *The Spider*, resplendent in her blue and white colour scheme, was the centre of attraction for

CAPT. C. D. BARNARD'S
Air Pageant

FLYING THRILLS ALL DAY.

PARACHUTE DESCENT
By Mr. JOHN TRANUM

THE AUTOGIRO
Passenger Flights for the first time.

"Daily Mail" Aviation Lesson.

AEROBATIC DISPLAYS
By famous pilots.

PASSENGER FLIGHTS
With CAPTAIN BARNARD
in his famous Monoplane—

THE SPIDER

[Accompanied by the Duchess of
Bedford, Captain Barnard made the
record flight in the Spider to India
and Capetown].

Mr. R. T. Proctor's Farm,
Wykeham, FRIDAY NEXT,

From 11 a.m. till dusk.

Admission 6d. Children 3d. Motor Vehicles 6d.

Programme and Music broadcast by "The Daily
Mail" super loud-speaker coach.

Left: The Air Pageant moves on to Spalding.
May 1931.

Below: Capt. Charles Barnard, fourth from the left,
and his team of pilots in Spalding, May 1931.
(*Spalding Guardian*)

Competition time. Win a free flight in *The Spider*, Spalding 1931. (*Lincolnshire Free Press*)

a crowd that grew as the afternoon wore on. The first flight for a ten-year-old Spalding schoolboy made a great impression on him:

> We sat side by side in wicker seats, like in a bus; only they don't need to be so comfortable because there is more to look at out of the windows. There was some cotton wool in little tins, like at the barbers [for ear plugs] and two clock faces that said 'speed' and 'height'. On the cabin wall there were a lot of picture postcards of Capt Barnard and his aeroplanes.
>
> When the propeller started to buzz round, everybody seemed to be looking at us through the windows. The whole cabin was throbbing like a bus does when it gets halfway up a steep hill and as we began to jolt across the field there was a bit of a smell of petrol. It became smoother in a little while and the man behind me said: 'we're off the ground!' I looked out of the window and the wheel near me was not going round any more. The ground was going away from us then it tilted and swung underneath. Below, I could see tiny people, other aeroplanes, the ruins of the abbey and the brown-coloured river. Some men were ploughing ever such straight lines in the fields that look very neat, with roads looking like a network of joined-up ropes.
>
> Now there was a lot of mist [passing through cloud] and then I saw our shadow on a cloud below so we must have been up a good height. Surfleet bridge and the church could be seen as we followed the river Glen to the railway bridge; turning over the green patchwork of

Record-breaking Fokker FVII, G-EBTS, *The Spider*, the main attraction in C.D. Barnard's air display programme. (R. Humm)

the golf course. When the aeroplane dropped downwards it felt like being in a lift and the engine made a noise like heavy rain on a tin roof. A greenhouse caught the sun and dazzled me as we passed. Then the field appeared again; we were down with hardly a bump and my first aeroplane trip was all over.

Appealing to both the patriotic and social consciences of his audience, Charles Barnard made it known that it was his intention to continue the 'National Aerodrome Movement' theme begun by Sir Alan Cobham in 1929. He declared that he would, however, strengthen its popular appeal by 'Bringing the thrills and lessons of the Hendon Air Pageants to people in all parts of the country. One of the most popular attractions will be to offer the public, flights at reasonable prices.'

Before the fare-paying public, however, came the inaugural free flight for VIPs of the local council. Councillors J.G. Bowditch; G.W. Chatterton; H.G. Frost; T.H. Padley and E. Dalton took advantage of this rare opportunity for a bird's-eye-view of the territory over which they presided.

It is interesting to contrast Cobham and Barnard; the former directing his earlier efforts quite specifically at councils, espousing the future for a network of regional aviation guided by his own expertise. He had a genuine interest, too, in the development of aviation on a much broader front, and to this end devoted much of his 1929 tour to taking a reputed 10,000 schoolchildren into the air. Cobham saw the future of aviation as a young science in the hands of young people. Barnard, on the other hand, spotted air shows as an immediate commercial commodity. Free trips for councillors was merely

Capt. Barnard prepares to fly Spalding Urban District councillors, May 1931. (*Lincolnshire Free Press*)

a good PR move, but to be fair he also espoused the involvement of youth. Publicity material issued to local newspapers stated his tour objective to include 'the provision of a scholarship fund for which Capt. Barnard, at the *Daily Mail*'s suggestion, is setting aside some of the profits to help local enthusiasts to learn to fly'.

While being laudable, there is a certain vagueness about it and no evidence of the fortunate beneficiary or beneficiaries has yet come to light. It would not be the last time such high-minded social objectives were put forward to the public as a good reason to attend air shows. The 1933 British Hospitals Air Pageant expressed similar but equally vague, aims. In his defence though, Charles Barnard's air show itself crammed in much entertainment between his arrival at Wykeham at 11 a.m. and dusk.

Throughout the day under a glorious sunny sky, while Charles Barnard kept *The Spider* busy flying ten passengers at a time in airliner comfort, Captain Edward Ayre offered exciting joyrides, in the Spartan Three Seater, including aerobatics for the more daring. Putting his scarlet-painted Avian through its paces with several aerobatic exhibitions and 'crazy-flying', Mr H.L. Stace kept the watching crowd enthralled with loops, rolls, side slips and stalls. With its short take-off and landings, in the hands of Reggie Brie the autogyro amazed the crowd, some of whom had a rare chance to sample its performance from the back seat.

At 4.30 p.m. a novel item, billed as 'The *Daily Mail* Aviation Lesson', began. Pupils and masters of Spalding Grammar School were invited to attend this set piece at the close of the school day. Arriving by bus they formed a group around the publicity vehicle, whereupon Mr William Courtney, the *Mail*'s aviation representative, began an enlightening

Reginald Brie became the first to qualify for a British helicopter pilot licence. (Crown)

talk on the principles of flight. Radio communication was set up between the ground and Capt. Ayre flying overhead in the Spartan Three Seater. Wearing headphones, the pilot performed a series of manoeuvres according to instructions received from Mr Courtney on the ground. In turn, these instructions were relayed to the crowd by means of loudspeakers attached to the publicity bus. Later the Spartan was joined by the Cierva autogyro, which gave an exciting demonstration of its unusual capabilities. All in all this was a good publicity stunt but one with a sensible, practical objective too; what boy could fail to be attentive during such a 'lesson'!

Closing the show, John Tranum, a Dane considered by many to be the most famous and skilled parachutist of his time, 'walked the wings' of the Spartan but because the wind was by now quite strong and blustery, his parachute descent had to be cancelled.

Via shows in Grantham and Lincoln, Capt. Barnard took his Air Pageant to Heacham near Hunstanton on 16 July at a place known as Stoney Hills with a programme similar to that seen in Spalding, where adults were charged 1s (5p) entrance, children 6d (2½p) and cars also 1s. Flights were priced at 5s (25p) for about ten minutes. Spalding must have been considered a good venue because, unusually, he put on another show there in October.

The year 1932 saw a veritable explosion of air shows upon the UK aviation scene. Charles Barnard, having set the pace during the previous year, took his tour to South

Upper image: Clustered around autogiro G-AAYP, pupils of Spalding Grammar School listen intently to Reggie Brie, 8 May 1931.

Lower image: Cierva C19 Autogiro, G-AAYP. (*Lincolnshire Free Press*)

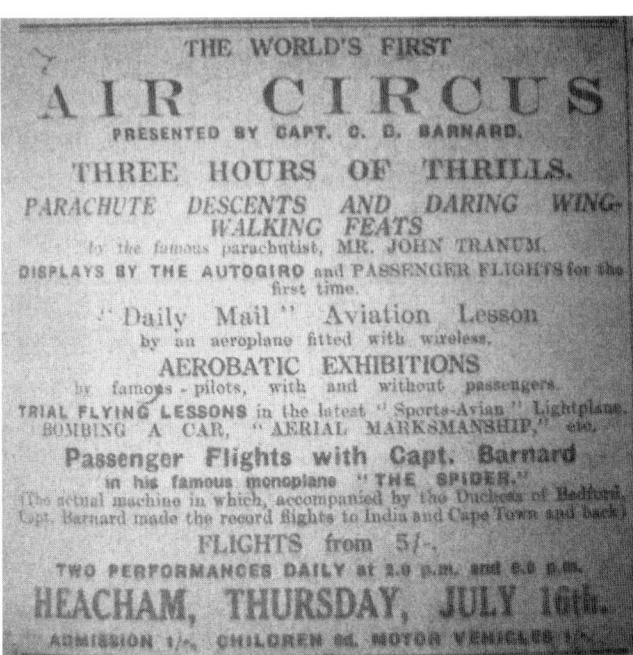

Capt. C.D. Barnard's air circus comes to town. Hunstanton, July 1931.

Africa in December 1931, then came back to England to rejoin the 1932 air show circuit but only as a 'featured' pilot rather than with his own team, probably because he would have found it impossible to compete financially with the extravagant, lavish scale of Cobham's own 'National Aviation Day' shows. While Barnard was on his UK tour, Sir Alan undertook planning for his own show but also spent a considerable amount of time that year on a route proving flight to Africa for Imperial Airways. After much careful logistical and financial planning – a Cobham hallmark – he began his own tour in April 1932. According to Harald Penrose in his book *British Aviation – Widening Horizons*, Barnard took himself off to South Africa in 1933 and then went off to India where he set up a touring air show in 1934, but his name faded from the scene thereafter. He was reported as joining the RAF in 1940 and ironically joined Cobham's Flight Refuelling Co. in 1942 until he retired in 1944.

Before giving a flavour of what Alan Cobham's air shows were like, it is worth contemplating the sheer scale of his operation. Such an enterprise not only needed logistical organisation but naturally needed financial backing too; Cobham had obtained this from Sir Charles Wakefield, who made his fortune in the lubricating oil industry.

Cobham's first show campaign ran from 12 April to 16 October 1932 and embraced 170 towns throughout the land. 200,000 joyriding passengers were carried aloft with only one accident and one fatality in that time. The magnitude of the undertaking and Cobham's commitment can be judged from the fact that by the end of the following year (1933) his air circus had completed a cumulative total of 1,216 displays in 506 towns. Twenty-two pilots, supported by 100 ground staff, had flown no less than 380,000 passengers in the two years. Is it any wonder, then, that even today there are thousands who can say 'I went to see Cobham' or 'I flew with Cobham'. Such was his charisma, that even those who in fact did not actually attend a Cobham show really believed they had done so – if it was an air show, then it must be Cobham's! Some of the pilots employed by Barnard also began to appear in the list of other shows, among them Reggie Brie with his autogyro and L.H. Stace.

Although private flying training was expanding, with flying clubs springing up all over the country, there was still a relatively small number of private pilots with 'A' licences who were experienced enough to undertake exhibition flying with reasonable skill and safety. Hence a scrutiny of press reports suggests a 'pool' of the leading exponents of the day being drawn upon for events such as air shows or air races, two major components of the 1930s air scene. A number of serving RAF pilots also participated in these events in a private capacity. There is no better example of this sudden release of aerial energy; fuelled by the efforts of men like Barnard and Cobham and ignited by headliners such as Amy Johnson and Jim Mollison; than when the seaside town of Skegness went air show-crazy in 1932.

Flying for entertainment of the public had, of course, been seen in Skegness before the First World War, but it was in June 1927 that pleasure flying established a foothold in the town, when Capt. Henderson rented the Camp Ground for the purpose of offering pleasure trips by aeroplane for the remainder of that holiday season. He also negotiated for a possible regular air service across The Wash at a similar site at Hunstanton, but did not follow it up. Around 1930, though, from a field on Roman Bank in Winthorpe, Michael

D.L. Scott and Capt. George Pennington began offering joyrides and flying lessons in DH60 Moth, G-AAKM and DH Puss Moth, G-AAXL, before they, too, succumbed to the lure of crossing The Wash by air on a regular basis.

Operating what was probably the first scheduled UK internal air service – the three-seater DH Puss Moth G-AAXL across The Wash between Skegness and Hunstanton – in the summer of 1931 Pennington and Scott carved out their own aviation niche under the banner of Eastern Air Transport Ltd. To eke out a living, the Puss Moth was kept busy by offering joy rides to holidaymakers from Winthorpe airfield. Soon, flying training was added and it was not long (1 April 1932) before the Skegness and East Lincs Aero Club opened its doors to budding pilots and boasted eighty-seven members by the date of the first air show, of whom twenty-seven were learning to fly. An early success was local boy Alex Henshaw, who became the first owner-pilot of the club to qualify for his licence. It is a matter of history, now, that Alex became one of Great Britain's most famous pilots, destined to find fame as an air-racer, air record breaker and a test pilot whose name is forever synonymous with the Spitfire and Lancaster.

For some months the owners of Skegness airfield had been in discussion with Mr R Richardson of Nottingham about operating a regular air passenger service from Skegness to Nottingham (Tollerton) airfield. Initially the DH80A Puss Moth, G-AAXL, would be used and the date set for the inaugural flight was Whit Monday 1932. Due to a clash with the Skegness Pageant scheduled for that same date, the first flight was delayed to the following day, Tuesday 17 May 1932. The flight actually started from Tollerton when the Puss Moth, piloted by Michael Scott, took off at 10.30a.m carrying the Lord Mayor of Nottingham, (Alderman W. Green) and the Lady Mayoress – a full load! Thirty-two minutes later, after an uneventful flight, they landed at Winthorpe airfield where members

Alex Henshaw, a founder member of Skegness Aero Club. (John Ketteringham)

of the town council and Capt. Pennington welcomed them. It was intended to extend the passenger air service during the summer months to Mablethorpe, Sutton-on-sea and across The Wash to Hunstanton. In view of the requirement to cater for joyriding flights as well, it was expected that a DH Fox Moth might need to be brought in to augment these services; even a Spartan Three Seater was being considered.

It was on 26 May that a flight across The Wash (a twice-daily service priced at £1 return) caused a rapid and dramatic reduction in the little airline's 'fleet'. Michael Scott, accompanied by Lewis Tindall, a flying instructor at the Skegness Club, was in DH 80A Puss Moth, G-AAXL heading for Hunstanton when, halfway across, an engine con-rod broke and smashed the cylinder head. Scott force-landed safely on a sand bank just as the sun was setting but more worryingly, just as the tide was rising. They could see Skegness quite clearly and indeed it looked so close that they thought someone would soon see their predicament and come to rescue them. No such luck.

Darkness fell without any sign of rescue so, with water lapping round them, they undressed and started to swim for shore. After an hour they had made absolutely no progress towards their target because the tide was pushing them further away. With difficulty they regained the sand bank but by now their clothes had floated away and the water was closing in on the aeroplane. In desperation Scott decided to set fire to the Puss Moth but with the loss of their clothes they had also lost cigarettes and matches. What to do now? Scott was clearly a good man to have in a tight corner. With a cool piece of lateral thinking he dipped a rag into the fuel tank, disconnected a plug lead and tried to set fire to the rag, by one of the men pulling over the propreller while the other shorted

DH Puss Moth, G-AAXL, force landed in The Wash on 26 May 1932. (Peter Green)

the lead onto the petrol-soaked cloth. They did this until, exhausted and just about all in, with bleeding hands and the water up to their waists, miraculously the rag caught fire and they pushed it into the petrol tank.

The resultant blaze lit the sky and was spotted by the crew of the cockle boat *Lizzie Anne* of Spalding. In the nick of time both men were rescued, very much the worse for wear and suffering from the effects of exposure – of both types! G-AAXL though, was no more and next day only the engine, propeller and a few small metal parts remained as salvage. The replacement was a DH Fox Moth, G-ABVJ, which continued to operate this and other duties until the end of 1935 when Eastern Air Transport ceased trading and sold its aircraft.

Winthorpe airfield and Skegness Aero Club, with its holiday atmosphere, catering and fuelling facilities, soon became a popular destination for aviator visitors but sadly the accident to AAXL was not the last serious incident.

The first air fatality occurred on 23 April 1933 when DH 60 Moth X, G-EBTV was being flown so low over the sea that a wing tip touched the water and the aircraft cartwheeled into the sea 150 yards off Seacroft Esplanade. Pilot Walter Shaw was rescued only due to the selfless efforts of Skegness man Victor Mastin and John Littlewood from Leicester who, fully clothed, swam out to the aeroplane and pulled him out. Sadly they were unable to save the passenger, James Barton, who drowned in the wreckage. For their courage that day, the two men were awarded Royal Humane Society Testimonials later that year.

'Skegness Plane Tragedy', 'Terrible Tragedy at Skegness', 'Air Tragedy' shouted the headlines in several local papers as they reported the tragic death of pilot Herbert Chantrey, aged twenty, and his passenger Miss Vera Field, aged thirty, in the crash of Avro Sports Avian, G-ABDN, on Tuesday 6 June 1933. Herbert was an engineer by profession; a keen sportsman who raced motorcycles and who learned to fly in 1932. He used an aeroplane frequently, often flying between London and Birmingham on business, and on the Saturday prior to the accident he and Miss Field, a long-standing friend of the family, left London to fly to Edinburgh where they met friends and members of the Motor Cycling Club and watched some motorcycle trials. On Sunday they flew down to Kirkby Moorside, near York, staying there until Tuesday morning when they set out for Heston aerodrome, London.

Glorious sunshine bathed Skegness as the Avian touched down at 1.30 p.m. in good time for a spot of lunch and topping up with fuel. It would no doubt seem an idyllic end to what was later said to be a holiday for Miss Field. Ground staff put twenty-five gallons of petrol in the tank and an hour later the couple were ready to go. Herbert made sure Vera was securely strapped in, settled himself in his own open cockpit and the engine was started. Club staff pilot James Hill, one of the few eyewitnesses, described what happened next.

The pilot made a good take-off into wind but at a height of 100 feet he straightened out and brought the aeroplane round into a left hand turn down wind, with hardly any bank. The machine went into a spin, made about a turn and a half, plunged into the ground and exploded in flames. If one got into a spin at one hundred feet there were a hundred chances to one of [against] getting out of it.

The aeroplane crashed into an adjoining field, owned by Mr R.J. Tagg, but intense fire prevented rescuers reaching the couple, for whom nothing more could be done. The spot was just fifty yards from the place where another, less serious, aero mishap had occurred only the previous Sunday.

Fortunately many decades were to pass before anything as tragic as this involving a civilian aeroplane at Skegness occurred again. On that sad occasion, though, Richard Goodwin and Ian Mailer, both employees of Skegness Air Taxis Ltd, lost their lives in a Cessna 150, which crashed 250 yards from the terminal building on Ingoldmells airfield in the evening of Tuesday 17 July 1990.

Eastern Air Transport left the scene around 1934, and although the aerodrome carried on under the management of Aircraft Distributors Ltd, which ran a flying school there for a time and Crilly Airways Ltd reopened the Skegness to Nottingham route for a few months in mid-1935, that company folded the following year, so that by 1938 the airfield at Winthorpe was in limbo. The country was also sliding towards war so it was unlikely that the situation would change in the foreseeable future. Lack of active commercial airfield facilities at the Winthorpe site, therefore, may have contributed to the demise of a very famous Percival Gull Six, G-ADZO, in a way that had remarkable similarity to the Puss Moth AAXL incident.

Mr H.L. Brook was a well-known record-breaking flyer, the contemporary and equal of any of the flying celebrities of the 1930s. For example, he captured the Cape Town to London solo record in G-ADZO on 5 May 1937 with a time of four days and eighteen minutes, which stood until February 1939 when Alex Henshaw, clocking thirty-nine and a half hours, blew a massive hole in Brook's record. He was well used, therefore, to the risks of flying, yet had a brush with the Grim Reaper in the very aircraft flown first by Amy Johnson then by himself on numerous record flights and attempts.

For a change, Mr Brook was making a simple business flight from Leamington to Norwich on 26 May and intended to break the journey at Skegness. He said he was unable to land at Skegness and so continued on across The Wash. He had been airborne for forty-five minutes, cruising at 1,000ft when at 11.15 a.m., well out over the sea, the engine faltered. One petrol tank had run dry so he switched over to another but despite switching the interfeed taps on and off, the engine would just not pick up. He commented:

> I saw it was hopeless … I was losing height rapidly and had insufficient height to get back to land. The only place to get down was on a sand bank behind me and I managed to land safely, coming to a halt at the water's edge. I could see a fishing boat nearby so I took off my shoes and socks and paddled out to it. By the time we got back to the aeroplane the water was up to the wings.

A choked fuel feed pipe had forced him down on Roger Sand in The Wash about eighteen miles out from King's Lynn. He was rescued by Ted 'Turks' Pratt, skipper of the cockle boat *Stagshead*, whose crew waited for the tide to rise sufficiently for the boat to float up to the stricken aeroplane then fixed ropes to tow it into Lynn. Progress under tow was so slow that the skipper feared he would not make port in time to discharge his valuable catch – of shellfish – and get it on the trains! Faced with the loss of a great deal of money he slipped the tow, anchored the aircraft and marked its location with a buoy.

The next day the crew of the *Stagshead*, with the assistance of two more boats, managed to tow their unusual catch back to King's Lynn. Two tides, though, had done a lot of damage to the Gull. Apart from having been totally immersed and battered by the sea, the undercarriage had been washed away and the wheels were buried in the sand. It was winched onto Common Staithe quay where it remained overnight, guarded by a policeman to keep souvenir hunters at bay. A dejected Mr Brook inspected the sorry spectacle before the wings were removed and this famous and once-beautiful aeroplane was taken away on a lorry to Croydon Airport.

Nevertheless, back in 1932, these were events in the unknown future. The scene was set for a vintage year of air spectaculars at this little seaside resort, beginning with a show put on by the Aero Club on Whitsunday 15 May 1932 – and what a cracking show it was too!

According to the *Boston Guardian*, ' ... somewhere approaching 20,000 people were thrilled by the air pageant. Of these, 15,000 paid for admission while the remainder viewed the proceedings from vantage points on the roads bordering the airfield.' (Sounds familiar!)

A major attraction was not only the variety of aeroplanes – no less than thirty-five were present – and pilots, but also the prospect of several exciting air races. Dull weather early on brought the cancellation of the longest event, a race to Nottingham and back. Later in the day the sky cleared sufficiently for a handicap race, of three fifteen-mile triangular laps, to take place. Six competitors entered the first heat, which was won by

Percival Gull, G-ADZO, sitting forlornly on a sand bank in The Wash, 26 May 1938. (*Lynn News & Advertiser*)

Visitor to the Skegness Whit Sunday 1932 air show, DH Puss Moth, G-AAXX, being refuelled from petrol pumps that served cars on the Roman Bank road and aeroplanes on the airfield. (Alec Davis via David Robinson)

Cierva C19 Mk III autogyro, G-AAYP, flown by Reginald Brie. From four entries in heat two, the Imperial Airways pilot Capt. Gordon Store flying Spartan Three Seater I, G-ABTT, emerged winner. In a tribute to the skill of the handicapper, there was little to separate the first three home in the final. With an average speed of 102mph Gordon Store crossed the line just two seconds ahead of Reg Brie, to win a splendid silver trophy, while Carill Napier, in Widgeon G-AADE, came third, a mere one second behind the autogyro. Prior to the racing, Avro chief test pilot H.A. Brown had put the prototype Avro Cadet G-ABRS through its aerobatic paces for the crowd in the first public outing for this brand new aeroplane. ABRS was then entered in the races, and in the final came home in fourth place with Roy Dobson at the controls.

An 'arrival landing' competition grabbed spectators' attentions, with much wagering as to whose wheels would stop closest to a tape laid out on the landing run – without the aid of engine or braking. In a remarkable demonstration of good judgement (or was it luck?) Flt Lt Christopher Clarkson, one of several RAF pilots participating privately in the pageant, managed to roll to a halt a mere fifteen inches from the tape.

Among the well-known pilots, many of whom had a slot in the programme, the only woman at that time to have won the King's Cup air race, Miss Winnie Brown in 1930, opened the proceedings with a spirited display in her all-black Avro Avian, G-EBVZ. Mr L.H. Stace, who will be remembered from the earlier Barnard shows, performed aerobatics in a Spartan Three Seater, being followed by Reg Brie in similar vein in autogyro G-AAYP. Fg Off. H.H. Leech, a member of the 1931 Schneider Trophy team, 'gave a wonderful demonstration of high speed flying, rolling, vertical turns and inverted gymnastics which won hearty plaudits'. RAF pilot Flt Lt Chris Clarkson followed this up with an exhibition of inverted flying in a DH Gipsy Moth.

No air show would be complete, of course, without a parachutist and on this occasion Ivor H. Price, who would shortly join Alan Cobham's team – only to lose his life while on

SPEED THRILLS SKEGNESS AIR PAGEANT

RING STUNTS

Record Crowds

...ching twenty thou-
...thrilled at the Skeg-
...held at the resort on
...f these, about 15,000
...dmission—the others
...dings from vantage
...aways around the
...s a great success,
...unfortunate that the
...Notting ...an had to be
...low clouds and poor

...pointment, both for
...and general public,
...ulation, because, as it
...ull weather of the
...s later gave way for
...the sun actually put

...having to be can-
...splendid programme,
...us had the oppor-
...m occurs of seeing
...t of various types.
...est-known types were
...ell-known pilots took
...lots being as under:
...Tayton and Mr. Ivor
...is, Lt.-Commander
...L. Flt.-Lt. Clarkson,
...Mr. Shipside, Mr.
...Presland, Mr. H. R.
...t, Mr. Hansel, Mr.
...J. Grierson; Aviax,
...Miss Winnie Brown,
...ce, Mr. Cohen; Puss
...Mr. Scholes, Mr. A.
...y Moth, Mr. Isher-
...f, Squadron-Leader
...ome; Arrow Active,
...Avro Cadet, Mr.
...r Lascelles, Civilian
...ke; Spartan, Col.
...r. Napier; Autogyro,
...t, Gibbins.

EXHIBITIONS.

...ened with a clever
...Miss Winnie Brown,
...ing's Cup race in
...is the only woman
...his trophy. She
...famous all-black
...being the identical
...she won the cup,
...ry high speeds and
...ghts, then indulging
...se dives and spins,
...making a perfect

Here we have a fine picture of the autogyro, with which some splendid displays were given by one of the test pilots of the Autogyro Company. This machine created a good deal of interest.

Mr. H. Price, who made a sensational parachute descent, still smiling after an experience few of us would relish.

The first major air display at Skegness on 21 May 1932. Upper image: Reggie Brie and Autogyro G-AAYP. (*Spalding Guardian*)

tour in 1935 – made the 'sensational descent.' Broadcasting a commentary on the whole proceedings was William Courtney, who brought along the *Daily Mail* 'Marconi-Coach' with its splendid loudspeaker system.

In those less regulated times, air racing was a popular attraction for pilots and spectators alike but it was not without its risks, which of course was part of the attraction anyway. Spartan Three Seater, G-ABTT, is a case in point. Whereas most of its contemporaries were used by joy riding enterprises, ABTT is believed to be the only one of its type that was raced regularly at that time. Exactly one week after the Skegness race, G-ABTT crashed during the *Morning Post* Air Race in which thirty aeroplanes competed in a race from Heston to Norwich and back. It was during the return leg that the Spartan flew into a tree at Stanton, near Bury St Edmunds, killing Flt Lt Frank Gibbons DFC, a former First World War pilot and well-known air racer, who was born and educated in Peterborough.

This Whitsunday event was meant to be a curtain raiser for the next and grandest of the three air shows planned for Skegness in 1932. Sir Alan Cobham's show had taken

Tense moments as a Spartan Three Seater approaches the winning line. Air racing at Skegness, 1932. (Alec Davis via David Robinson)

to the skies in April, and Tuesday 12 July was scheduled to be 'National Aviation Day' in Skegness.

In the event the show was a bit of a disappointment – not through any fault of Cobham but due to the vagaries of British weather. Low cloud and poor visibility on the morning of the show prevented most of the Cobham flying entourage from leaving Goole for Skegness and by 3 p.m. only five pilots had managed to complete the transfer. Although the civic reception was cancelled, with Sir Alan having to settle for lunch in the airfield clubhouse, the excitement was in fact about to begin.

Having already suggested the lavish style of Cobham's show, this is now evidenced by the arrival of a large twin-engine, fifteen-seater Handley Page biplane airliner – with something of a bang! Described as being 'on hire from Imperial Airways for joy-rides', the airliner is believed to be a Handley Page W8b, G–EBBI.

As it crossed the Skegness threshold, one of the airliner's wings struck a parked car with a resounding blow, gouging a large hole in the car roof and smashing its windscreen. Fortunately for the owner and his companion they were strolling blissfully unaware in the public enclosure at the time or would surely have been decapitated! In losing a chunk from its wing, the W-8 failed to put down inside the airfield but wobbled on to make a safe landing some distance away in the general direction of Burgh le Marsh. Enough to give Sir Alan a little indigestion, no doubt!

From Grantham on 14 June 1932, Sir Alan's National Aviation Day moved on to Boston then Doncaster before being given the full civic treatment by Mayor Councillor W.F. Elderkin on Bracebridge Heath airfield, Lincoln, on 17 June. Reports of Cobham shows in Lincoln newspapers suffer from a sad lack of detail about the programme and over the years tend to concentrate more on who got the benefit of a free flight!

This reportage is in marked contrast to their contemporaries writing for the *Boston Guardian*, *Lynn Advertiser* and *Cambridge Times*, which are much more fulsome. On this occasion, apart from the whole of the city council – as usual – the *Echo* reporter described his own trip in a DH Gipsy Moth flown by Mr Bebb. For the demanding researcher, though, he did redeem himself a little with the following interesting snippet. The word 'LINCOLN' had been painted across the flat top of a new gas holder in Bracebridge. The sign measured 26ft by 14ft, with letters of 3ft strokes – said by a sister paper to be readable at a distance of one mile:

> Tribute to Lincoln Corporation's initiative in having the name of the city painted on the new gasometer recently erected at the Bracebridge gas works, was paid by the pilots of the fleet of aircraft which visited the city today with Sir Alan Cobham, the famous aviator. The general manager of the display, Mr D.L. Eskell, said 'It's a fine idea and it will be very helpful to pilots, especially in fog. I don't know another city where this has been done and this afternoon I intend to go up to ascertain the height at which this name can be read. I shall pass this information straight on to *Flight* and *Aviation* so they can advise pilots.'

If a pilot could see this particular unlit sign in fog, it seems highly likely that he or she was about to have a very nasty accident with said gasometer!

In contrast to the refusal he received on his Municipal Airports Tour of 1929, Sir Alan persuaded the Air Ministry to allow him to use Westwood Aircraft Park, Peterborough, for a show on Tuesday 19 July 1932. Sir Alan attended in person but that was not always the case, as his main role was to generate business and organise the general conduct of the campaign. Lady Cobham, who ran the London base operation but frequently accompanied her husband, on this occasion telegraphed her apologies to her Rotary Club hosts.

Crack of dawn saw his entourage pouring into Westwood Air Park in a stream of thirty cream and emerald painted cars and lorries, so that the morning performance could begin promptly. Sir Alan arrived at 1.30 p.m. by train from London and met up with the civic party, lead by the mayor, Councillor J. Mansfield, at a splendid lunch in the Grand Hotel.

Local newspapers were always supplied with pre-show press releases but it was civic lunches and dinners that gave Sir Alan access to a captive audience of influential local government grandees and businessmen who made prime targets – if not always compliant when it came down to raw financial detail – for his aviation mantra. We need to bear in mind that, like a missionary, he was the first person to reach out to the ordinary people of Great Britain and tell them about his vision of and for the future of aviation, as it would affect individuals and local communities. Yes, of course it would be good for Alan Cobham, too – but why not?

With military precision, tents for pilots, ground staff, catering, administration and tradestands, were soon erected on the grass airfield and the sound system of Kolster-Brandes loudspeakers was strung out along the crowd line. The only hiccup was that, despite advance publicity suggesting twenty aircraft would be on display, only eleven arrived from the previous day's show held in Market Harborough. However, in an effort to broaden the show's appeal to the public, a Leyland Hippo low-loader lorry brought *Golden Arrow*, the racing car in which Sir Henry Seagrave established a new world land speed record at Daytona Beach on 11 March 1932.

The *Peterborough Advertiser* ran a coupon campaign for twenty-five free flight tickets in the weeks running up to the show. Each lucky winner was photographed on completion of the trip and received a souvenir card containing a photo of Sir Alan, his personal message and a picture of the aircraft in which they had flown.

With a running commentary by Dallas Eskell, the tour manager, three flying programmes filled morning, afternoon and evening and despite poor attendance during the morning, what was described as 'a comfortably large' crowd for the afternoon performance swelled to 'enormous' proportions in the evening, lining the ropes ten and twelve deep in some places. Of course the inevitable additional thousands of 'free' spectators took up vantage points in hedgerows and lanes all around the airfield. Like B.C. Hucks before him, this made Sir Alan hopping mad so he devised a bizarre deterrent in the form of a motorbike and sidecar, driven around the lanes bordering the show fields like a maniac by one of his ground staff!

Opening with a grand flypast by all aircraft, there were no delays between each act as the crowd was thrilled to every style of aerobatics conceivable, including air-racing round pylons and parachute drops. Balloon-bursting was very popular and involved the pilot Flt Lt Charles Turner-Hughes diving down on small inflated balloons while leaning out of the cockpit apparently firing shots at them with a revolver! He was only firing blanks, though, and the balloons were actually burst by a hidden marksman with an air rifle on the ground. The comedy car chase with flour bombing was the real thing though. One act, guaranteed always to take the crowd's breath away, was wing walking. Martin Hearne climbed out of the cockpit and calmly walked along the lower wing of an Avro 504, then sat with his legs wrapped round one of the undercarriage wheels while the pilot looped the loop. From there he rounded off his daredevil act by climbing onto the top wing where he crouched on the centre section, like a surf rider, holding on to a small piece of rope in the teeth of the 90mph slipstream. Even launching the glider caused much excitement when it was towed across the field at 25mph behind a big Chrysler sedan, before releasing its tow rope at a height of 400ft to include a loop and several circuits of the airfield and then slipping silently and gracefully to a perfect landing. During the morning session, the mayor, members of Peterborough Corporation and members of Oundle Urban Council returned bubbling with delight from their free opportunity to gaze upon the sunlit city from the comfort of canvas armchairs in the Airspeed Ferry, G-ABSI, *Youth of Britain II*. While the mayor was aloft the mayoress, evidently a little nervous, was heard to remark, 'Before I go up I want to see what the mayor looks like when he comes down!' But they all trooped off to the grand lunch so she missed her chance. Brisk business by the ten-seat joy riding aircraft and the smaller Avros, including two of Cornish Aviation Company's pillar-box red machines, continued all day until dusk, with hundreds of paying visitors, from four-year-old Harold Laxton to eighty-five year-old Mrs Ada Potter, having made their first trip aloft. The show was the hottest gossip in the city for many days afterwards.

Cobham's Airspeed Ferry, G-ABSI, gives councillors an opportunity to see their towns from above. (Cobham Plc archive)

An almost constant procession of aircraft took passengers aloft, for this was where the real money was made, and one of the best aircraft around for that purpose was the ubiquitous Avro 504. This sturdy and reliable machine had been in civil and military service in one form or another since 1913. During the 1932 season Cobham hired some of Cornwall Aviation Company's remaining Avro 504Ks, among them G-EBIZ and AAAF, to ensure the joy riding public did not suffer delays from too few aircraft or unserviceability. When the tour split into two teams during 1933, Cobham augmented the existing joy riding element with more Avros from North British Aviation Co., which itself acquired 504s from the now defunct Northern Air Transport Company. By 1934 he had used fourteen 504s at various times but mishaps gradually eroded his small fleet until at the close of that year's programme only G-EASF – the oldest 504K then flying – was left. At the end of 1935 the Air Ministry refused to renew certificates of airworthiness for rotary engine aircraft, which brought the era of the Avro 504 to a close.

Having thrilled thousands in Peterborough, Sir Alan motored off to a meeting in Cambridge while his circus moved on to Heacham (Norfolk) the next day, 20 July 1932, where, amongst the hurly-burly of the show, there was an amusing sequel to that less than happy landing by the W8 at Skegness.

While the display got under way in Heacham, Sir Alan, having rejoined the show after his meeting, flew across The Wash to Skegness to see how repairs to his Handley Page were progressing. Mr A. Dunseath, an accountant from King's Lynn, went along for the ride in Cobham's Tiger Moth. The flight was uneventful, repairs were proceeding well and Sir Alan prepared to depart once more for Heacham when a scenario unfolded that was reminiscent of a 'Laurel & Hardy' film so popular in those days!

Another stalwart Avro 504K, G-EBIZ, gave sterling joyriding service for Cobham's shows.

Acquired by Cobham, former Berkshire Aviation Avro, G-EASF, was the oldest 504K still flying in 1935. (G. Stuart Leslie)

Swinging the propeller of the Tiger Moth, its engine opened up with a roar (our trusty knight had not set the throttle with his usual care) and the aeroplane immediately began to move forward at quite a pace. Stepping smartly out of its path, Sir Alan ran round the wing and made a grab for a wingtip as it trundled past him. This in turn caused the wayward beast to swing round in a circle – at the same time Mr Dunseath was trying to catch hold of the other wing. For a few scary moments the latter became pursued by the aeroplane until he, too, ran out of its path. Having eluded these two aviators the Tiger now set its sights on the Handley Page, careering towards the airliner as if determined to make its acquaintance. Fortunately its path took it clear of the W-8, but only by a whisker was a collision avoided. In the end, the Tiger's wheels hit a pile of wood causing it to pitch onto its nose, whereupon the propeller broke, the engine stopped and sanity returned to the scene. Undaunted, our intrepid knight telephoned his colleagues in Heacham for a spare propeller to be flown over 'immediately!' No one argued when Sir Alan gave an order, and the new prop quickly arrived. Two hours later he took to the air, wisely following a coastal route back to Heacham.

Despite heavy showers, eleven aircraft put on a stirring show, similar in almost all respects to that of Peterborough, from a large grass field on Church Farm, Stoney Hill, between the villages of Heacham and Ringstead.

While Sir Alan and his entourage toured other East Anglian towns, Skegness and the holiday-making public prepared itself for yet another air pageant laid on by the local aero club. Playing its part in the pre-show hype, the *Boston Guardian* informed readers that:

Aerial novelties never before witnessed in this country, will form special features of the programme at the Skegness and East Lincs Aero Club's Air Pageant, which will be opened at 2.30pm on Sunday July 31 [1932] by Miss Amy Johnson. She will be accompanied by her fiancé, Mr J A Mollison, now that the latter's projected double crossing of the Atlantic has been postponed until September. Mr Mollison will bring to Skegness the identical machine on which he broke the England to Cape record earlier in the year.

Graphic descriptions of the aerial acts scheduled for the programme were written, followed with the final promise that 'the Skegness Air Pageant will be the finest show to be seen outside London'.

Closed circuit air racing, where competitors would remain in sight of spectators, was to be a further major attraction for the paying public. The latter had a choice of paying 1s 3d (6½p) for the general public enclosure, or 2s 6d (12½p) for entry to a special enclosure. To put these prices into perspective, in 1932 when an agricultural worker's wage was about 35s (£1.75) a week, 1s 3d equates to about £3 in 2008.

Wretched weather right up to the start of the show put a damper on attendance but even so a large crowd gathered bubbling with expectation. The biggest disappointment, though, was the non-appearance of Amy Johnson and Jim Mollison, who were to have flown in specially to open the proceedings. Since Amy and Jim, whose names were linked amorously as well as professionally, had 'tied the knot' on 29 July, just two days previously, it is perhaps understandable why they might have been 'otherwise engaged'! Four years were to elapse before the famous duo actually managed to fit in a flying visit to Skeggie

GREET
AMY JOHNSON
AND
J. A. MOLLISON
AT THE
SKEGNESS AIR PAGEANT
Sunday, July 31st, 1932

THRILLS NEVER SEEN BEFORE.

'Planes hurtling to earth in a
thrilling " Dive " Race.
Three Parachutists falling sheer
with unopened Parachutes.
Autogiro engaging in aerial combat
with a Gipsy Moth.

OTHER SPECIAL FEATURES.

Aerobatics on high-speed machines.
Two Handicap air Races.
Motor Cyclist's ride through burning building
and plate glass sheet
Blind and inverted flying.
Demonstrations of new type fighting and
commercial 'planes.
Motor Cycle football match.

SEE THE FINEST DISPLAY IN THE MIDLANDS.

GATES OPEN 11-3J A.M.   Pageant Commences 2-30 p.m.
PRICES (including tax)
PUBLIC ENCLOSURE 1/3. SPECIAL ENCLOSURE 2/6
AMPLE PARKING FOR CARS.

The main attractions, Amy Johnson and Jim Mollison, failed to appear at the July 1932
Skegness air show.

sion of their second annual sports and tea.

FAMOUS FLYING COUPLE VISIT SKEGNESS

Popular Jim and Amy Mollison visited Butlin's new Skegness Holiday Camp at Whitsun week-end, where they were given a most enthusiastic welcome. Here they are seen with Mr. Billy Butlin (on the left), Mrs. Butlin (between her husband and Mr. Mollison) and Mr. Butlin's little daughter Shirley.

Amy Johnson and Jim Mollison finally make it to Skegness in 1936. Seen here with holiday camp owner Billy Butlin. (*Spalding Guardian*)

and their celebrity status was still powerful enough for most of the region's newspapers to report the event:

> Mrs Amy Mollison, the popular air heroine, paid her first visit to Lincolnshire on Whit Sunday, May 31 1936, when accompanied by her husband Jim she flew to Skegness from Heston. The couple spent several hours at the new Butlin's holiday camp, located on the sea front between Skegness and Ingoldmells, as guest of Mr Billy Butlin and his wife and daughter Shirley.

The weather throughout the morning had been vile, alternating showers of rain and hail being supplemented by a cold, gusty wind but just prior to the arrival of the Mollisons the sun came out and almost summer-like conditions prevailed. Arriving at Winthorpe airfield in a hired DH Puss Moth piloted by Amy, she and Jim were mobbed by a crowd of some thousands and given a rousing civic welcome by members of Skegness Council lead by Mr Fred Cooper. When Amy made a short speech of thanks she said she had had to dodge some storms en route, and when asked for her opinion of Skegness as a holiday resort, she replied tactfully, 'I am beginning to think I have missed a lot of things through not paying a visit to Lincolnshire before now.' That evening they flew out for an engagement in Bridlington.

Ford Trimotor, G-ABEF taxies in at Skegness in July 1932. (Peter Green)

Refuelling Ford Trimotor G-ABEF by hand at Winthorpe, Skegness, July 1932. (Peter Green)

To return to that show in 1932, Pennington and Scott excelled themselves and the splendid flying programme coupled with an opportunity for the public to move freely among the thirty aeroplanes and pilots, helped make the day a resounding success. Great interest was shown in the largest aeroplane present, an all-metal, fifteen-seat Ford Trimotor monoplane, G-ABEF. It was demonstrated by Major F. Digby DSO but was not used to carry passengers that day. Flying events opened with an impressive formation flypast of all thirty visiting aeroplanes, lead by aircraft flown by members of the Skegness Aero Club.

Six competitors, of whom Alex Henshaw in his Gipsy Moth was one, took part in the opening thirty mile handicap air race. Consisting of three ten-mile laps, the race was

Comper Swift, G-ABWW, piloted by Flt Lt Nick Comper, ready to race at Skegness in 1932. (Alec Davis via David Robinson)

won by Col Louis Strange in a Spartan Three Seater II, G-ABTR, at an average speed of 100mph, and by the narrow margin of three seconds over Capt. A.J. Styran. This was an excellent result for the Spartan ABTR, for it was the prototype Mk II and had been sent out on to the UK racing scene for publicity purposes.

Skegness to Nottingham (Tollerton) was the course for the main race of the day, a total distance of 120 miles. From the four entrants, Col Strange again emerged the victor, this time by ten seconds over Fg Off. H.H. Leech, with Flt Lt Nicholas Comper third, another twenty seconds behind. Once again the handicapper did an excellent job, contributing much to the level of excitement in the closing stages of each race.

Rounding off the day's proceedings, Capt. Pennington organised a social evening in the clubhouse. Not everyone, however, was convinced it was a good idea. In contrast to modern safety regulations, the view of many in those more informal days was that drinking and flying were two activities that need not necessarily inhibit each other. Skegness magistrates, however, took a dim view of this and an application to the Bench, for an extended drinking licence for a dance to follow the air show, brought a terse response from its chairman. Pleading his case (with a little name-dropping thrown in) Capt. Pennington pointed out that:

> The Prince of Wales' pilot and other noted pilots are coming to the event. It would give a
> bad impression if they could not get a drink at Skegness when they could get one on similar
> occasions anywhere else in Europe.

The chairman of the Bench still remained unimpressed and rather brusquely rejected his application!

As mentioned above, in a large field atop one of the few Norfolk hills overlooking The Wash on 20 July 1932, Alan Cobham treated the Heacham district to one of his vintage shows, a feat which he would nevertheless surpass the following year in Boston. On 10 September, Skegness rounded off its vintage year of air shows with the third of the Aero Club's flying meetings at its Roman Bank airfield. The main event of this meeting was a handicap race from Skegness to Boston Stump and back. The five entrants included three Comper CLA7 Swift single seat racers, designed by Flt Lt Nicholas Comper and built by his company. Comper himself flew over from Liverpool (135 miles in forty-nine minutes) to take part in the race, coming second in the Gipsy-engine G-ABWW behind another Swift.

The *Boston Guardian's* reporting of Cobham's display of 31 May 1933 can be regarded as one of the finest and most descriptive of any such air show accounts found in the course of nearly three decades of research. Its style and content are reminiscent of that same newspaper's account of the great aviator B.C. Hucks's first visit to the town in 1912, leading one to wonder if it was the same anonymous reporter still employed some twenty years later.

Under the banner headline 'GUARDIAN READERS FLY OVER BOSTON', the reporter's enthusiasm matched that of most first time fliers. Dutifully he accompanied the mayor, Cllr C.W. Fleet and members of the council on the ritual inaugural flight (free of course!) of the day. Having flown before, on the occasion of Cobham's visit to Boston the previous year, he was obviously seeing this latest visit through the eyes of an 'experienced' air traveller.

Although described yet again as 'a large airliner', the photograph printed with the report can be clearly identified as that of a Handley Page W10, believed this year to be G-EBMM.

In contrast to my previous flight there was on this occasion a haze which obscured more distant objects; for example The Wash being only dimly visible through the murk. More

Handley Page W10, G-EBMM, served Cobham well. Seen at air shows all over the Fens in 1933. (R. Humm)

directly below, however, the Stump [St Botolph's church]; streets; the river and shipping in Boston dock could all be seen clearly, my fellow passengers vying with each other to spot places of interest.

What probably impressed my colleagues most was our return to the flying ground in Boardsides, off Sleaford Road. We passed over it and the airman flew low, only to regain height quickly, giving the uninitiated the sensation which one receives when suddenly ascending in a fast elevator. It was a surprise but I think everyone enjoyed it. In the cabin were two dials informing us that the aeroplane travelled at 80mph at a height of 800 feet for the trip and I estimate we covered about twenty miles.

Upon his return to terra firma Mayor Fleet waxed lyrical about the joy of flying while Sir Alan made his usual stirring speech about his campaign, the impending boom in aviation and how an aerodrome would bring untold benefit to the town. Some years would elapse before an airfield was in fact established in Boston but nothing on the scale envisaged by Sir Alan.

During the preceding week the *Boston Guardian* ran a 'guess the height' competition, with free flight tickets as prizes. This entailed readers submitting a coupon on which they wrote their estimate of the height at which an aeroplane would begin trailing smoke when it flew above Boston as a preliminary to the show. And indeed the Wednesday preceding the show saw 'a bright yellow, high-speed, single seat fighter, of a type until recently used by the crack squadrons of the RAF' flying over Boston. Following a spirited aerobatic display it then emitted the expected trail of red smoke. Estimates of the height,

A rare flying shot of SE5A, G-EBVB, used for skywriting and pictured at Alan Cobham's air show in Wisbech, June 1933. (Via David Rayner)

submitted by hundreds of entrants, ranged from 114ft to no less than 10,700ft! It might be said that both the former and the latter person may have needed an optician more than a flight! In fact the altitude was 645ft and the nearest twenty-five entries won complimentary 10s (50p) flight vouchers. Twelve more lucky winners were selected from the audience of the Scala cinema who participated in yet another competition during Tuesday evening's performance.

Almost certainly the aforementioned aeroplane was a First World War vintage SE5a. In 1922 a number of this type were purchased by Major Jack Savage for the purpose of 'skywriting' – an early form of aerial advertising. Many were the occasions in those days – well before the bite of modern-day environmental and safety regulations – when the public, on hearing the rise and fall of aero-engine noise, would raise its eyes to the sky and behold the word VIM, PERSIL or BUICK appearing in coloured smoke as if by magic. Except for the removal of guns and headrest and the extension of the exhaust pipes to the full length of the fuselage, these SE5a aircraft were in all respects standard First World War fighters in civvies. The exhaust pipes joined in a Y-shape to the rear of the rudder, and when a smoke-producing substance was introduced by the pilot from a tank inside the fuselage, controlled by a lever in the cockpit, lots of lovely, coloured smoke belched rearwards.

Photographic confirmation of this type of aeroplane appearing in Cobham's shows has been found from June 1933 when G-EBVB put in an appearance at the Wisbech (Leverington) event. That particular SE5a was scrapped the following year but because Jack Savage carried on using these aeroplanes – and more importantly, maintained them – well into the 1930s, at least three ex-Savage SE5s have survived the ravages of time and are now preserved for posterity in the RAF Museum; at the Shuttleworth Trust, Old Warden, and the Science Museum, London.

Those lucky ticket winners provided our *Boston Guardian* reporter with many column-inches of material and five post-flight letters describing the experience were also printed. From these it can be deduced that winners went aloft in two batches in the Handley Page airliner, which was fitted out with twenty comfortable armchair seats. Their trips lasted fifteen minutes and on at least one of the flights it was remarked upon that the airliner was in close formation with other aeroplanes. One winner, Mr R. Cabourn of Carrington, was obviously a keen flyer and wrote:

> The airliner is a safe means of travelling quickly for young and old, as evidenced by one of my fellow passengers, a lady of seventy-six years of age [Elizabeth Loftus]. For those who want a thrill, though, the smaller planes give this opportunity. Later in the evening I went in the air race, in a three-seater piloted by Capt Miller and we reached 105mph.
>
> My first experience of flying was in September 1929 when I had an ordinary passenger flight over Boston with Western Aviation Ltd, since then I have been keenly interested. In June last year I had an aerobatic flight in one of Sir Alan Cobham's planes, also at Boston. Nothing would give me a greater thrill or more pleasure than the opportunity to become a qualified air pilot.

At the other extreme, Miss Mary Royce, a pupil of Boston Girls High School, composed her own impressions in a rather more breathless style:

What a thrill! A free ride in the multi-engined air liner *Youth of New Zealand* [G-EBMM]. A double thrill when a mechanic invites me to sit beside the pilot in the cockpit. One almost has to crawl through the tiny door to reach the seat by the pilot and immediately one's eye is attracted by the numerous dials on the dashboard.

The propellers whirl and with almost the same movement as that of the driver of a car the pilot starts the machine and we bump along the ground to face the wind.

The pilot pulls the joystick towards him; the machine rises and the instruments on the dashboard begin to move. To me it seemed that we were going into a hedge but with a quick movement of the joystick, we miss it and are soon flying over the town, which looks like a toy village.

As we circled the Stump the pilot leaned out of the cockpit and I wondered which instrument I should grab if he disappeared overboard! The Witham, flowing at the side of the Stump, seemed no more than a dyke and it seemed impossible that this was the same river on which I was boating last Saturday. The town looked deserted but as we approached the landing ground, people below looked like flies. We passed over the field to get into the right position for landing and as our wheels bumped along the ground it seemed as if we might collide with another aeroplane just taking off. But no, the flight is over all too soon and I am left looking forward to one day flying my own machine and seeing Boston as a pilot not a passenger.

Come to the air show! HP W10 G-EBMM leads the flypast formation. (Cobham plc archive)

On the strength of so many glowing testimonies, certainly Alan Cobham could be well satisfied that his expressed aim of making the British public air-minded was achieving great success.

Although attendance in the afternoon was not as large as it would have been on a market day, a tremendous crowd assembled for the evening show. It was treated to the works. Formation flying; aerobatics to make even the brave duck their heads; inverted flying; flour-bombing a car; closed-circuit racing and parachuting, together with continuous joyrides in the Handley Page and any other aircraft that happened to be available. It was all vintage Cobham!

The beginning of the decade had seen no let-up in record-breaking attempts in the aviation world. 1931 had been a vintage year among many and it was the year in which the Englishman, Charles William Anderson Scott, made his mark in aviation circles. This former sailor, boxer, QANTAS bush-pilot and RAF fighter pilot took off from Lympne aerodrome on 1 April and flew solo to Australia in a DH60M Gipsy Moth in nine days. Not content with this he left Australia on 26 May for the return trip to England, which he completed in ten and a half days. In recognition of his holding the record for both directions at the same time, Charles Scott was awarded an Air Force Cross.

This was only the first of his record attempts and he, too, decided to cash in on his popularity by becoming associated with an air show company called British Hospitals Air Pageants Ltd (BHAP) which went on the road on 8 April 1933. On the same day, Wednesday 31 May, that Alan Cobham was wowing the crowds in Boston, Charles Scott, touring under the banner of Sky Devils Air Circus, brought his show to Waddington Aerodrome, Lincoln. Scott and his (1932) record-breaking DH 60M Gipsy Moth, VH-UQA, were stars of the show, which opened in glorious weather with a flypast over the County Hospital and round the cathedral several times before returning to Waddington for the pageant to swing into action.

Fourteen aeroplanes were billed in the programme, including a DH 84 Dragon and a Fairey Fox I. Among the pilots were well-known names from the record-chasing fraternity, including the Hon. Mrs Victor Bruce, Capt. R.H. McIntosh of Imperial Airways and Col James Fitzmaurice, former commandant of the Irish Free State Army Air Corps. The two latter gentlemen came to prominence in 1927 when, as a crew, they made an unsuccessful attempt to become the first to fly the North Atlantic from east to west. It is an interesting coincidence that the aeroplane in which they made that attempt was none other than Fokker FVIIa G-EBTS *Princess Xenia* which, re-named *The Spider*, you will recall began this chapter in the hands of Charles Barnard.

To illustrate further that research is not always quite as straightforward as simply reading press reports, the wording of the *Lincolnshire Echo* report for this show is a little confusing in respect of the DH Dragon. The aircraft in the display were said to 'include a 14-seater Handley Page airliner of the very latest type and identical with the machine that Mr and Mrs J A Mollison are using in their proposed flight to America in the near future'. Amy and Jim Mollison were not going to use a Handley Page aeroplane but were indeed planning to use a DH 84 Dragon for their Atlantic flight on 8 June – only a week after this very show. The Dragon could be configured, at most, for only six to eight passengers – not fourteen. On the other hand a Handley Page airliner, like the W8 for example, could carry twelve to fourteen passengers. Which one was it at this air show, then?

Well, BHAP did in fact buy a Handley Page W8b from SABENA (OO-AHJ/G-ACDO) in December 1932, with the intention of using it for joy riding the following season, but it remained in a dismantled condition for two years and was scrapped before ever coming into service. Only days after Scott's show was due to leave Lincoln for its next venue in Boston, Alan Cobham's team was due to arrive in Lincoln on Wednesday 5 June. Now, as we saw earlier, Cobham actually used a Handley Page W10 in his show and it had been photographed while he was in Boston. The *Echo*'s report of events at Lincoln began with an account of the BHAP show then went on to 'trail' Cobham's impending visit, so it seems quite likely that the reporter simply mixed up his description of who had what aeroplane. By way of further confirmation there is also photographic evidence that BHAP used a DH Dragon when the show visited Spalding on 3 August.

The Fairey Fox I was quite a potent machine and its appearance on the programme follows a general pattern of beefing up these shows with at least one high performance aircraft. Originally produced as a two-seat military day bomber, only three examples appeared on the civil register and BHAP hired one of these, G-ACAS, for joyriding. At Lincoln it was flown by Mrs Victor Bruce, and who better to make the day memorable for fourteen-year-old Joan Twidle from Wragby, visiting the show with her mother, as one of the winners of a free flight competition? She greatly enjoyed her flight over the city in the 200mph open cockpit Fox and would remember it for a very long time. A great deal

Programme from Charles W. Scott's British Hospitals Air Pageant at King's Lynn in 1933.
(Ray Wilson)

longer, in fact, than that particular Fox would remain flying, because it caught fire in the air and burned out upon landing just six weeks later!

Passengers were carried in the aeroplanes taking part in the afternoon; evening flying displays and joy riding continued until dusk. Then it was off to Boston the next day (1 June) to repeat the programme to another audience. BHAP was back in the region at King's Lynn on 2 August, flying from Sayers Marsh, Exton Place, with the West Norfolk and King's Lynn General Hospital designated to be a beneficiary of some of the proceeds. The next day Scott brought the show to Spalding (Thursday 3 August 1933). This, its 110th venue, was held in Proctor's Willow Tree field in the shadow of Wykeham Abbey, just a stone's throw from the location of Cobham's 1929 show.

Held in what was described by the *Lincolnshire Free Press* as 'ideal weather', Scott brought a dozen aeroplanes to entertain the large crowd. Stars of the show were undoubtedly the DH60 Gipsy Moth, VH-UQA, in which he flew to Australia, and DH 84 Dragon airliner, G-ACCR, both of which were used for joyriding on the day. Other aeroplanes known to be at this show are Avro 504K G-EBYW, DH83 Fox Moth G-ACCF, Spartan Three Seater G-ABKK, an unidentified Monospar, and a machine described as a Desoutter cabin monoplane. Led by Charles Scott the flying team included Pauline Gower in her green Sparton KK, *Helen of Troy* (reported as 'still feeling the effects of a minor mishap yesterday'); Flt Lt J. Pugh; Fg Off. Gerald Hill; Capt. A.L. Robinson; Mr L. Anderson; Jock Bonner; Capt. Earl B. Fielden; Mr J.K. Morton; Mr G.E. Watson; Mr C. Longmore and Capt. W.A. Rollason. One of the other notable

Aircraft parked at Wykeham, Spalding, for the British Hospitals' Air Pageant of 3 August 1933. (*Lincolnshire Free Press*)

characters listed among the pilots was Capt. Percival Phillips DFC MM. This dashing forty-year-old Cornishman, a former First World War fighter pilot, was one of the first to enter the barnstorming business way back in the early 1920s, founding the Cornwall Aviation Company, whose bright red Avro 504 aeroplanes became synonymous with that era. By the 1930s his reputation was so great that Sir Alan Cobham sought and used his experience, and later Percival became a director of Charles Scott's enterprise. It is sad to record therefore, particularly in view of his vast experience of the Avro 504 aeroplane, that his exuberant streak finally got the better of him on 13 February 1938 when he was killed while flying his own Avro 504N, G-ACRE. Doing a spot of 'racing' at low level while en route to a friend's house in Gamlingay (Cambridgeshire), he misjudged hopping over a small wood – or simply did not see it – and crashed into a tree.

The usual aerobatics – balloon-bursting, paper-cutting, flour bombing, parachute descents, closed circuit air races and continuous pleasure flying – went to make up a very lively programme of events at Spalding. In those days the paying public was freely allowed to fly as passengers with the display pilots, during even the wildest gyrations, something no longer permitted in this modern age. Admission to the ground was 1s 3d (6½p) for adults, 6d (2½p) for children and 1s (5p) for cars. Flights were priced from 4s (20p). Spalding resident Hartley Egar was one of the lucky free flight winners and he described his experiences in a letter to the *Lincolnshire Free Press*:

> I made the journey from the town to the flying field at Wykeham by tandem and sidecar. When I arrived I met up with another free flight winner and we decided to go up together. Hearing over the loudspeakers that bookings were being taken for the 'Formation Flight' we tried to get a seat but were too late. However, we were given the option of flying in a 'cabin' or an 'open' aeroplane and decided that we should see a great deal more from an open cockpit, so chose the latter. Tickets were issued to us for the Rollason Desoutter 'Open' biplane. [In his excitement it seems likely that Mr Egar became confused about his aeroplane since the Desoutter was a cabin monoplane].
>
> Our pilot was E W 'Jock' Bonner and as we walked to the machine I noticed he was wearing his old college cap. Clambering into the cockpit, my friend took the front position, [most likely in an Avro 504 biplane with its two-place rear cockpit] the engine revved and we were soon in the air. The machine climbed to about 400 feet and the pilot executed an acute left hand turn, which made us hold on as we were not strapped in [!] We flew as far as Spalding railway station and on the return journey another aeroplane carrying lady passengers came alongside until it seems as if the machines were touching. After the pilots had exchanged conversation for a short time we made for the ground, swooping down gracefully to a perfect landing.

Miss E.M. Holland, a young lady from Whaplode, was another winner of a free flight ticket and she took hers in a General Aircraft Monospar twin-engine, four-seat, cabin monoplane, with Mr J.K. Morton as pilot.

Then another day brought another venue. Friday 4 August saw the show at Long Eaton (Nottinghamshire), Saturday it was in Derby, Sunday in Liverpool, and Monday

"FLYING FLEA" AT SKEGNESS

Our composite block shows a group of Skegness officials with M. Mignet (extreme right), and below the inventor is seen with his machine. "It is impossible for it to get into a spin, it is not an aeroplane but a kite and a parachute," he said.

Monsieur Henri Mignet was a big attraction at Skegness airfield and Cobham's Boston show in September 1935.

Opposite above: Cobham's air circus comes to Lincoln. His advertisements took several styles.

Opposite below: Cobham's air circus at Sutton–on–sea.

SIR ALAN COBHAM'S
Great New AIR DISPLAY

MIGNET'S
FLYING FLEA!
IS COMING!

SIR ALAN COBHAM'S AIR DISPLAY

BRITAIN'S GREATEST
AIR SPECTACLE!

**Sleaford Road,
BOSTON,
TUESDAY, SEPT. 17th,**
ONE DAY ONLY.
Continuous from 2.30 p.m.
till dusk.
Admission 1/3, Children 6d.
Cars 1/-, Cycles 2d.

FLIGHTS FROM 4/-.

See the New Autogiro
and JOAN MEAKIN, the
Glider Girl!
12 New Aircraft in 20 Daring
New Feats!

Mignet's Flying Flea gets top billing at Cobham's show in Boston in September 1935.

in Llandudno. By the end of their UK tour – which would coincide with the end of 'summer time' in early October – the organisers, BHAP, said they hoped to have visited 200 places and raised thousands of pounds for hospitals all over the country. They let it be known that 10 per cent of the takings on this occasion would, for example, be donated to Johnson Hospital, Spalding.

It was Alan Cobham's shows that held the sustained interest and support of the British public year after year. There was almost no corner of the country that his team did not visit and often revisit several times. Between 1932 and 1935 Heacham, for example, was venue for four Cobham shows, Boston and Lincoln three times, while Skegness, Sutton-on-sea, Peterborough and Wisbech were each visited twice.

Cobham's National Air Day arrived in Lincoln from Boston on Wednesday 7 June 1933, and in contrast to BHAP, used the old Bracebridge Heath airfield as its venue. Not to be confused with RAF Waddington, the former was a grass airfield previously used by the now defunct Air Ministry Aircraft Acceptance Park, and in 2007 terms was located behind the large hangar and sheds still to be found adjacent to the A15 road just inside the city limit, running up to the boundary of the old Mental Hospital site.

It has to be said that press coverage of any air show held in Lincoln has been found to be pretty thin on detail and column-inches, and this one was no exception. There is nothing to indicate that anything exceptional took place, nor are there any names of participating pilots or aircraft – except, of course, Alderman J.W. Rayment, the mayor who, with members of the corporation, was flown in the inaugural free trip, and Sir Benjamin Bromhead of Thurlby Hall who, at the ripe old age of ninety-five made his first flight, together with his great-grandson.

Sir Alan though, chose this venue to introduce another of his bright ideas on how to squeeze more productivity out of the principal revenue-earners – i.e. the joyriding aircraft. A brand new 800-gallon, six-wheeler petrol tanker joined the show for the first time. Brimming with National Benzole petrol the idea was to speed up refuelling and aircraft turnaround by having the tanker move around the field to each aeroplane whenever it required refuelling. The aeroplane was filled up where it stopped, thus doing away with the need for it to taxi across the airfield to a fixed refuelling point and back every time it needed topping up. This trimmed off a substantial amount of time during the course of a show and allowed many more fare paying passengers to be flown.

Ely, too, had a return visit and a lengthy article by an enthusiastic reporter in the *Cambridgeshire Times* captured the atmosphere of the all-action display on 11 September 1934. This year also saw a return to a single team format with just eight aeroplanes led by the mighty, twenty-seater HP W10 G-EBMM Youth of New Zealand staging 'undoubtedly the most wonderful aerial exhibition ever seen in this district'. The report continued:

> Thousands of people flocked to Capt R G M Wilson's field on the Little Downham road, with many hundreds more lining the roadside for a free view. From the time of the arrival of the formation of eight aeroplanes, until the last flight when dusk fell, this troupe of intrepid airmen delighted and thrilled both the large crowd and the large numbers who made flights in one or another of the machines.

To set the record straight though, it was not only airmen since, in addition to the powered aeroplanes, there was also a glider on the programme and an accomplished lady pilot, Miss Joan Meakin, flew it with equal panache. It was noted that Sir Alan did not attend the Ely show as he was 'engaged on important work for the Air Ministry'.

The programme opened with a grand formation flight and air salute by all the aircraft, filled with the first batches of fare-paying passengers. The formation was lead by the HP W10 flown by Sqn Ldr C.H. Bembridge. Immediately after take-off the smaller aircraft formed up into a vee on either side of the big airliner and cruised sedately around the locality. Back over the field the formation changed to line astern, with each aeroplane in turn flying low along the crowd line and dipping a wing to salute the Civil Air League ensign.

Billed as 'The Famous Three Aces', C.H.W. Bebb, T. Jones and 'Daredevil' Jock Mackay, launched into a synchronised precision aerobatic display in three Avro Cadet biplanes, one

Airspeed Ferry, G-ABSJ, at a Cobham air show in Wisbech in June 1933. (Via David Rayner)

Clambering into an Avro 504 with decorum was not easy for a lady in a skirt! Cobham's show at Wisbech, June 1933. (Via David Rayner)

The three aces. The red, white and blue Avro Cadet trio that performed formation aerobatics at Cobham's shows. (Cobham plc archive)

coloured red (G-ACOZ) and the others white (G-ACPB) and blue (G-ACLU). In tight formation manoeuvres, these civilian pilots showed the public that anything the RAF could do they could do equally well. Capt. Bebb must have indeed impressed someone, as it is recorded that for a time during the Spanish Civil War he was General Franco's personal pilot.

The fact that the show had reverted to a single-team format with the somewhat reduced number of participating aeroplanes compared with previous years, could be seen as a sign of the plateau of the popularity of air shows as entertainment. This is not to suggest a marked change or even a decline in Cobham's popularity – that was never in question for the public was far more interested in him than it was in the antics of Herr Hitler! It was more the strange result of being so successful! These air shows, just like their pre-First World War forerunners and even back to those far-off nineteenth-century ballooning exhibition days, set out to show the public that aviation was accessible to all and in this they succeeded – but to the extent that attendances waned while costs increased. As in the past, occasionally something was needed to restore a bit of spice to the programme to maintain gate receipts.

Speed and risk were the things most people wanted out of an aeroplane display and in order to beef up this spectacle, Cobham acquired a Blackburn Lincock for his show. This was a diminutive, purple-coloured, biplane, one of just three examples built as a possible lightweight fighter for the RAF. It was unsuccessful in that role and Cobham used the all-metal Mk II and III versions as aerobatic aircraft in his 1933 and 1934 tours, first appearing at his Skegness show in August 1933. It is interesting to note that the first Lincock, the wooden Mk I, G-EBVO, used in a similar way during 1931 and flown by A.M. Blake, was certainly displayed at an air show – not one of Cobham's though – held on a Lincolnshire beach, before being dismantled in August of that year. Geoffrey Tyson's precision solo aerobatics earned him acclaim throughout the aviation world and he was just the chap to put the Lincock through its paces. Changing mounts later in the programme, Tyson performed his remarkable, white-knuckle party piece of snatching a handkerchief from the ground, with a hook fixed to one lower wing tip of his Tiger Moth. Having interviewed Tyson later, a *Cambridgeshire Times* reporter enthused:

> The extreme accuracy of judgement and delicacy of touch that the feat demanded will be appreciated when it is remembered that the wing tip is some fifteen feet to one side of the pilot's eyes. He has to concentrate not only on keeping his machine on a straight line from the wing tip to the handkerchief but also on keeping his wing tip only a few inches from the ground without actual contact. On the first three attempts he missed the handkerchief by an inch or so but on fourth attempt he was successful.

After Ivor Price had made a 'pull-off' parachute descent it was the turn of Joan Meakin, billed as the first English girl to become proficient in air-towed gliding. In her Rhönbussard sailplane, she was towed up to 1,500ft before release, whereupon she displayed the glider to perfection:

> … with a series of noiseless loops, rolls, spins, dives, turns and side slips until, with consummate ease she swept her frail craft across the flying field to alight with the grace of a bird.

Just how close the joy riders of Ely courted disaster in G-EBMM they will never know and it is all bound up with the reason why Sir Alan did not attend his Ely display personally on this occasion. He was very preoccupied with trials in preparation for an in-flight refuelling proving flight to Karachi, scheduled for 22 September, for which HP W10 G-EBMM was due to be detached from the air show programme to play a vital part. Fitted out with in-flight refuelling gear at Ford airfield, it would rendezvous with Sir Alan's Airspeed Courier G-ABXN off the Isle of Wight, top up his tanks air-to-air, thereby allowing him to fly non-stop to Malta. Over Malta, Cobham's second W-10, G-EBMR was positioned to air-refuel the Courier again so that he could carry on to Aboukir and Basra, where the RAF would provide air refuelling from two converted bombers. EBMM did its job entirely successfully but unbeknown to Cobham at the time the W10, having landed back at Ford to have the refuelling kit removed, crashed due to structural failure of the tailplane while in transit to rejoin the air show team at Coventry. All four people on board, including Capt. Bembridge, were killed.

Taking risks was all in a day's work to a rising new star that joined Cobham's team in 1935. The name and photo of this attractive young airwoman was splashed across newspapers and magazines all over the country. Naomi Heron-Maxwell, billed as 'The Society Girl Parachutist', was about to make her thrilling contribution to the air show scene.

Miss Helen Naomi Heron-Maxwell, second daughter of Sir Ivor Patrick Heron-Maxwell (eighth Baronet), was twenty-two years of age when she joined Cobham's team.

A star of Cobham air shows was the attractive society girl parachutist, Naomi Heron-Maxwell.
(Nicholas Heron-Maxwell)

She had learned to fly a light aeroplane a few years earlier and then turned her attention to parachuting. There is evidence in *The Sketch* magazine (10 October 1934) of Naomi making parachute drops on behalf of Charles Scott's British Hospitals' Air Pageant show in Woodford, Essex, in October 1934. In another issue of *The Sketch* (27 March 1935) Miss Heron-Maxwell was said to be 'studying gliding in Bavaria and plans to return to England in April [1935] to join Sir Alan Cobham's Aerial Circus as an amateur parachutist'. This was supported by *The Bystander* magazine (24 April 1935), which carried a striking portrait of this very attractive young lady, with the caption that she 'is appearing in a parachute act with Sir Alan Cobham's display, that starts its 1935 tour this month'. Final, graphic confirmation of Naomi's participation appeared in the form of a dramatic pair of photographs in *The Sketch* of 24 April 1935. One shows her clinging to a wing strut of Cobham's HP33 Clive airliner G-ABYX Astra while the other shows her at the instant of being pulled off the wing by her streaming parachute.

Naomi married Francis Allen in 1938 but sadly her husband died in 1939. Between February 1942 and October 1945, as First Officer Naomi Allen, she put her courage and flying skill at the disposal of the nation in the Second World War, becoming one of the longest-serving ferry pilots in the Air Transport Auxiliary (ATA). Naomi married Howard D. Thomas, an American, in 1957 and she died in the USA in the early 1990s.

Young ladies from the higher strata of society were expected, in those days, to behave with decorum so Joan Meakin and Naomi Heron-Maxwell presented Sir Alan's publicity department with a great opportunity to enhance their public persona in the pre-show press releases. In 1935 for example, Joan was billed as: 'The Pied Piper of the Air – who is calling upon the youth of Britain to follow her into the air.' While both young women were also described as:

> the reverse of the popular conception of the 'modern girl'. They do not smoke, never drink cocktails, do not enthuse over dress, dancing or 'dates' and dislike late nights. Their 'daring' is of another kind.

This, then, is background to the innocuous heading of a paragraph in the *Spalding Guardian* of 20 July 1935, entitled 'Parachute Demonstration':

> Everyone present was excited and thrilled by the pull-off parachute descent from the *Astra* airliner, by Miss Naomi Heron-Maxwell, 'The Parachute Girl'. She stood on the wing of the plane until it was several hundred feet from the ground, when she opened the parachute by pulling the rip-cord; it streamed from the case on her back, snatched her off the wing and floated her safely down to the ground, landing right in the middle of the airfield right in front of all the spectators. Miss Heron-Maxwell demonstrates parachuting as an amateur.

She most certainly was not an amateur.

This was all part of an air show held on Monday 15 July 1935, which was remembered with great enthusiasm by Ernest Brown of Spalding. Mr Brown experienced the joy of his first flight at the show when he was eleven years old. He recalled the field used was in May's Dyke Lane, Fleet Church End, a few miles from Holbeach. His trip cost 5s (25p)

FREE FLIGHTS FOR OUR READERS

A HORNCASTLE PAGEANT

VISIT OF SIR ALAN COBHAM

THRILLING SPECTACLE TO BE PRESENTED

On Wednesday, September 4th, Sir Alan Cobham is to present his great Air Display on the Louth Road, Horncastle. By arrangement with Sir Alan, "Standard" is able to place before readers an outstanding opportunity a free flight in one of the magnificent aeroplanes to be brought to Horncastle by the great airman.

In filling in the coupon below each readers has the chance of being among a happy bunch of who will be given free flights in Sir Alan's air liners.

Before has Sir Alan presented and entertainingly the lessons this National Aviation Day Campaign has great new display, which comprises 20 entirely new events and includes such famous figures as Miss Joan Meakin, the celebrated glider girl; Mr. Louis Rowley, one of the finest aerobatic pilots in the country; Capt. Idwell Jones, the latest "wizard" of crazy flying; and Mr. A. Harris, the most expert British parachutist. Two air liners will be seen in the display and amongst the new craft will be an autogiro and examples of the finest types of British civil craft.

NOVEL EXHIBITIONS.

Here are some of the remarkable exhibitions, all absolutely novel, which Sir Alan Cobham has arranged:—

Looping in formation. A new feat by daring pilots whose aeroplanes are tied together by ribbons from wing to wing. The cleverest exhibition of close formation flying yet seen.

Autogiro. A demonstration of the "wingless wonder," the most extraordinary aircraft in existence. The autogiro has no wings or control but the pilot controls the machine

Some of the 'planes that will visit Horncastle.

Miss Joan Meakin, the glider girl.

open, from Penzance to the Orkneys, and every town which possesses an aerodrome is being linked up in a thriving system of inland airways.

the great "Astra" air liner, with its Bristol "Jupiter" engines developing 1,100 h.p. and its spacious cabin seating 22 passengers. Flights are available from 4s.

THE NEW AUTOGIRO.

The demonstration of the new "direct control" Autogiro at Sir Alan Cobham's Display will be the first public exhibition of this wonderful new invention in Lincolnshire.

This new Autogiro has been described as the "wingless wonder," since it dispenses entirely with any kind of movable control surface. It has no elevators, rudders or ailerons.

In this machine, Senor Juan de la Cierva has made the most revolutionary advance in the method of controlling an aircraft since the Wright brothers first flew.

The pilot's sole control is a "joy-stick" connected to the shaft carrying the revolving "rotor." He steers the machine, banks it on turns, and causes it to climb or descend by inclining the "rotor."

In other respects besides control, the new Autogiro claims to be a great improvement on any other aircraft in existence.

It can take off with a run of only a few yards, and it can land almost vertically, with practically no forward run.

"SKY TRAFFIC COP."

Owing to its ability to fly slow

Glider girl Joan Meakin thrilled the crowds at Cobham's show in Horncastle, 4 September 1935.

– equivalent to about £10 nowadays – and was both a source of pleasure and pain to him. He had been pea-pulling that day and earned the sum of five shillings for his labours. Without telling his parents he went along to the flying field and blew the whole lot on a flight in the Astra G-ABYX. Ernest remembered every detail of that short trip around the Long Sutton area but winced and smiled wryly when he recalled his father's severe reaction against his backside afterwards!

According to the *Spalding Guardian*:

The thrilling programme began with an aeroplane 'paper-chase'. The pilot released a long paper streamer and whilst it was still falling through the air he endeavoured to cut it into small pieces by flying repeatedly through it. Next followed a demonstration by a Cierva

Naomi Heron-Maxwell making a pull-off parachute drop from the wing of Handley Page HP33 Clive I, G-ABYX. (Colin Cruddas)

Cierva C19 Mk IV, G-ABGB, was one of several autogiros used for Cobham's air shows over the years. (A.J. Jackson collection)

Autogiro, which can take off with a very short run; it can also hover in the air and, with a good wind blowing, actually fly backwards – at least in relation to the ground.

Although passengers were taken for 'flips' in all the planes taking part, the most interesting one as far as they were concerned was the formation flight and parade. Passengers were taken for a long cruise over the surrounding country and on returning to the airfield, the planes, following the *Astra* airliner and flying at top speed, each took their turn in a parade along the line of spectators.

Flight Lieutenant Geoffrey Tyson, the only man who has succeeded in flying upside down across the English Channel, gave a wonderful exhibition of high-speed aerobatics.

Nearly everyone present took part in the free flight competition arranged by the Ford Motor Company Ltd. Competitors had to judge the height of an aeroplane when the pilot discharged a smoke signal. The winner was Dr R E Crockett of Sutton Bridge, who correctly judged the height to be 957 feet.

An air race was flown over a course comprising three laps of a triangular route, the airfield being crossed on each lap. The three Avro Cadets took part and each pilot carried two passengers.

Aptly described as 'Flying's most difficult Feat' Flight Lieutenant Tyson's party piece of looping-the-loop through a hoop then picking up a handkerchief with a hook on his wingtip produced a real thrill. At the fifth attempt he caught it. One of the most interesting features – if not the most exciting – of the programme was a demonstration of gliding by George Collins, a member of the London Gliding Club. Mr Collins was towed behind an aeroplane by means of a wire rope to a height of 3,000 feet where he cast off. Then followed a series of loops, rolls, spins and inverted flying and a fast dive over the spectators – all without motive power.

Another free flight competition, this time much more difficult, was held. A plane flew over the airfield twice – the first time slowly, the second time much faster. Spectators had to judge the speed of the machine both times and subtract one from the other. Mr E Knight of the Talbot Hotel, Holbeach was the winner with a difference of 49mph.

An exhibition of inverted flying at twenty feet from the ground was given by Flight Lieutenant Tyson, followed by a humorous interlude, entitled 'Say it with Flour' [the ever-popular flour bombing routine] which caused roars of hearty laughter.

Throughout the display the *Astra* airliner made long journeys with passengers to places of interest in the district. Flown by Flight Lieutenant H C Johnson, chief pilot of the display, the *Astra* carried twenty-two passengers in its spacious cabin. Teas were served on the ground and afterwards the same programme was repeated.

The next day, Tuesday, all the machines, pilots and ground personnel journeyed to Bourne (Lincolnshire), where they gave another show to large crowds in a field off the Sleaford Road, Morton. Twelve more locations from Nottinghamshire, through Lincolnshire, Yorkshire and up to Newcastle were fitted in before Cobham's Astra Show returned once more to Hunstanton (Heacham).

The possibility of tragedy striking always stalks an air show, and to some extent it is in the very nature of the event. Despite such intense itineraries – at the rate of a show a day – and action-packed, loosely regulated programmes, there were thankfully few fatalities in

Handley Page HP33 Clive I, G-ABYX, *Youth of Australia.* The centrepiece of Cobham air shows in the mid-1930s. (Colin Cruddas)

the 1920s and 1930s. One such incident, however, marred the Cobham show in Ramsey (in Huntingdonshire in those days) on Tuesday 30 July 1935.

The Grunau Baby glider was at that time playing an important part in the embryo Luftwaffe pilot training scheme over in Germany but one example on the British register was flown regularly by George Collins, of whom we heard earlier and who was widely acknowledged as an expert glider pilot and current holder of the national distance record. George was aero-towed to 2,000ft where the rope was cast off and he began circling several times above the former First World War airfield of Upwood (it was not redeveloped as an RAF airfield until January 1937). Although the Baby was generally considered as semi-aerobatic, he rashly attempted a 'bunt' or outside loop, a manoeuvre for which the glider was not stressed. There was a sudden loud crack audible from the ground and the port wing of the glider broke away from the fuselage. Falling immediately into a spin, from which there was no chance of recovery, the glider dived into the ground at the edge of the airfield, killing Mr Collins instantly.

1935 was a relatively bad year for Cobham in this respect, as in March, John Tranum, one of his regular parachutists, had died in a coma after being taken ill at 25,000ft altitude during an attempt to break his own free-fall parachuting record. Within a few months of this loss, twenty-seven-year-old Ivor Price was killed in a parachuting accident during one of Cobham's National Aviation Day shows in Woodford. The worst incident of all, though, was a mid-air collision over Blackpool in September of that year, in which three people died. Tragic though they are, this relatively small loss of life should be set against an excellent safety record in the four years during which Cobham's touring enterprise carried almost one million people into the air.

Although deserving of his place in aviation history's hall of fame in his own right, Charles Scott's name will forever be linked with that of another contemporary pilot, Tom Campbell Black. Furthermore, both men are inseparable from the all-red DH88 Comet racer G-ACSS Grosvenor House in which they flew and won the 1934 MacRobertson Mildenhall to Melbourne air race. Like Scott before him, Campbell Black launched onto the air-show scene in 1935 teaming up with Lt Owen Cathcart-Jones, a former Fleet Air Arm pilot credited with making the first night landing on an aircraft carrier and – more dubiously – 'bombing' the flagship of the Med fleet with rolls of toilet paper! In that Royal Silver Jubilee year they intended to visit 180 venues and reached Green Lodge Farm, Oakham (Rutland) on 19 June, where the principal aeroplanes on show were Campbell Black's single engine Avro Cadet and a twin-engine General Aircraft Monospar (probably an ST-4 model) flown by Cathcart-Jones.

The pair arranged to bring their show to Wykeham on Friday 30 August, as part of Spalding's celebrations to honour the King. Disappointment, however, awaited spectators gathered near Capt. Proctor's Wykeham Abbey farm. With the amount of record attempts and air racing entries planned by both pilots in conjunction with their respective sponsors during the summer months of 1935, there was no way they could possibly find time to fit into the nomadic air show scene. The stars of the show conveyed their apologies but regrettably were unable to attend due to prior commitments.

In the case of Campbell Black, on 8 August he was winging his way in DH88 G-ADEF on yet another (successful) attempt on the England to Capetown record. Four days later he was back in England and deep in preparation for a series of long distance flights in the DH88, including yet another Cape attempt, this time in Mew Gull, G-AEKL, scheduled for September. Meanwhile, Cathcart-Jones was regularly involved on the air-racing circuit and his absence from Spalding was almost certainly due to preparations for the King's Cup air race on 6 September. Cathcart-Jones, too, had built a justifiable reputation as a record breaker, having many attempts and successes to his name, dating back to 1931. He had made numerous flights from England to the Cape and was co-pilot of another DH88 Comet racer (G-ACSR) that came fourth in the MacRobertson race. Another of his many claims to fame was that he had been a ferry pilot for the Nationalists in the Spanish Civil War.

Sadly, Tom Campbell Black was destined only to grace the air show scene for a few more weeks after the Spalding event. Aged just thirty-six, he died in an aeroplane collision accident on the ground during an air show at Liverpool airport on 19 September. Strange to relate, too, that Charles Scott's career also ended abruptly at the age of forty-three, when in 1946 he took his own life while serving with the United Nations in Germany. Owen Cathcart-Jones, meanwhile, continued to thrive upon his reputation in aviation and his name was even spotted in the credits of the 1942 Hollywood flying movie *Captains of the Clouds*, starring the legendary Jimmy Cagney. Cathcart-Jones had a small acting part playing a Canadian flying instructor and provided technical advice for the film – a role he repeated later the same year in *Desperate Journey*, another wartime flying film, starring Errol Flynn and Ronald Reagan. Evidently the lure of Hollywood or its climate must have suited Cathcart-Jones, as records show that he died in California in 1986.

In any event, for the Silver Jubilee show back in rural Spalding, it turned out to be typical English summer weather – cold and wet – but the names of the two (absent) aviation stars on the programme was still sufficient to draw a crowd, despite the inclement weather. A biting cold wind put paid to the parachute display by Harry Ward and W.G. Hure and only a few token flypasts were made by the half dozen light aeroplanes that were present. According to the local newspaper, though, the most notable events were the free flights distributed among winners of its competition and of course these could not be allowed to fall victim to the weather. Many column inches were devoted to one winner in particular, a Mrs E. Turner of Spalding who, at ninety-four years of age, was apparently the oldest person in the town to have flown in an aeroplane – piloted by a gentleman billed rather eccentrically as 'Stainless' Steele! All during the drive out to the temporary airfield at Wykeham Farm Mrs Turner regaled the *Lincolnshire Free Press* reporter with an account of her life and observations on the current state of the world and this seemed to make up for the lack of reporting of any other aerial happenings of note. This dear old lady also managed to grab the reporter when she landed, for he wrote:

> Her face aglow with pleasure and her voice ringing with enthusiasm she said 'Young man, you tell them that if there is another display, I shall only be too pleased to go up again. It was wonderful! I enjoyed it tremendously and I was not even a little bit afraid.'

With the Jubilee display failing to live up to expectations and going out like a damp squib, it was in some respects a reflection of the point which civil air shows had reached.

By the latter part of the decade the drawing power of civil air shows was in decline in much the same way that the balloon exhibitions had experienced fifty years earlier. Another factor was that with the rapid expansion of the RAF, Cobham lost many of his pilots and ground staff to the lure of the new Auxiliary Air Force and Volunteer Reserve and to commercial organisations involved with the now-burgeoning aviation industry. His National Aviation Day shows had run their course and he sold off his equipment and business – but not the business name – to Charles Scott. Scott operated an air show in 1936 but it seems to have faded away after that, perhaps because Scott himself was still getting involved with long distance flights. Cobham himself moved on to greater business ventures in aviation among which in-flight refuelling proved to be one of the most significant and long lasting. In the 1990s his company Flight Refuelling (FRL) changed its name to Cobham plc, which is a fitting tribute to his aeronautical vision.

Many years later, in a letter to Sir Alan, his former glider pilot Joan Meakin reminisced about her life with the air circuses:

> It was the excitement, freedom, comradeship and the sheer fun of it all that I adored, living the life of a gipsy, moving off each day to a different town – everyone keen and happy. I loved the bustle and noise and amusing chatter of ground engineers and gate staff as they worked. I loved the smell of the old Avros when the engines were being run and the sound

of the latest jazz tunes floating across the flying field from the loudspeakers. All these years later, were it possible, I would join the Display tomorrow to experience again the thrills of seeing Geoffrey Tyson flying upside down so low that the top of his rudder parted the long grass, or Jock Mackay crazy-flying or Martin Hearne with his fascinating wing walking – or all the wonderful things the pilots did to enthral the public.

Sir Alan, summing up his own view of those days, said, 'If I can claim to have done something useful with my life, this is chiefly because I did so much to bring so many people into contact with aviation while its heroic age still lasted.'

It was not that the general public was deprived of its dose of aviation thrills – there had always been the annual RAF Pageant and the SBAC industry shows at Hendon – but the world was changing and the civilian shows struggled to compete with government-funded events. As the dark clouds of war began to gather, it fell more upon the Royal Air Force to sustain the public appetite for aviation thrills. So, from 1934–39, as economic factors brought the decline of civilian travelling air circuses, annual Royal Air Force 'At Home' events were formally co-ordinated with the existing Empire Day (24 May) celebrations to introduce what became known as 'Empire Air Days'. These events were staged in late May each year at a selection of RAF stations across the UK, such as – in

What it was like to take part in a Cobham grand formation flypast, led here by HP33 Clive, G-ABYX, with Fox Moth G-ACEY and Avro 504K G-ABVH on its starboard wing.(Cobham plc archive)

ROYAL AIR FORCE STATION - - WITTERING

Empire Air Day   SATURDAY, MAY 20th, 1939

FLYING PROGRAMME—PART I

Event No.	Time p.m.	EVENT
1	2.45	**Formation Flying.**

The aircraft used in this event are Blenheim (Fighters) from No. 23 (F) Squadron and Hawker Hurricanes from No. 213 (F) Squadron. The Blenheims are two-seater fighters driven by two Bristol Mercury engines. Their speed is approximately 285 m.p.h. The Hurricanes have a speed of well over 300 m.p.h. and have a Rolls Royce Merlin engine. Their armament consists of eight machine guns per aircraft.

2	3.10	**Fast and Slow Flying.**

The public will be able to see the retractable undercarriages being raised and lowered and the differing speeds at which these aircraft can fly.

3	3.20	**Wireless Controlled Aerobatics—Gloster Gauntlet.**

The public will be able, for the sum of sixpence, to request the pilot to carry out such manœuvres as loops, turns, climbs, glides, rolls, rocket loops, etc., by radio telephony.

4	3.35	**Pupil's First Solo.**

Imagine yourself executing your first solo!—this is the pilot's impression of the result!

5	3.50	**Attack on Bombers by Modern Fighters**

Six Blenheims from No. 23 (F) Squadron representing Bombers will be attacked by six Hawker Hurricanes from No. 213 (F) Squadron.

6	4.10	**Individual Aerobatics—Hawker Fury.**

A Hawker Fury single seater Fighter will carry out a series of loops, upward rolls, slow rolls, half rolls, off loops, rocket loops, etc.

7	4.25	**Attack on Transport by Fairey Battles.**

A convoy of troops will be seen moving across the aerodrome Suddenly aircraft appear and attack the convoy with machine guns and bombs. The convoy stops, the troops scatter, and defend themselves with anti-aircraft machine guns.

Above and opposite: A typical programme for the last Empire Day air show before the Second World War, held at RAF Wittering on Saturday 20 May 1939.

ROYAL AIR FORCE STATION - - WITTERING

Empire Air Day   SATURDAY, MAY 20th, 1939

FLYING PROGRAMME—PART 2

Event No.	Time p.m.	EVENT
8	4.40	**The Shooting Gallery.**

A lesson in marksmanship from the air.

| 9 | 4.50 | **Section Drill—Hurricanes and Blenheims.** |

The aircraft will demonstrate air drill—an important part of a Squadron's normal routine. The sections will be seen changing into different formations.

| 10 | 5.0 | **Demonstration of Harvard.** |

The Harvard is an aircraft recently purchased from America for training purposes.

| 11 | 5.10 | **Individual Aerobatics—Gauntlet.** |

No. 213 (F) Squadron were previously equipped with this type.

| 12 | 5.20 | **Demonstration by No. 1 (City of Leicester) Squadron of Air Cadets.** |

| 13 | 5.25 | **Wing Drill—Fairey Battles.** |

A demonstration of formation flying by medium bombers.

| 14 | 5.40 | **Dog Fight—Fury and Gauntlet.** |

One Pilot from No. 23 (F) Squadron and another from No. 213 (F) Squadron will endeavour to out-manœuvre one another, each trying to "sit on the other's tail."

| 15 | 5.50 | **Wing Drill—23 and 213 Squadrons.** |

Wing drill by Modern Fighters as a contrast to Event No. 13.

| 16 | 6.10 | **Fly past and Demonstration of various Aircraft in the Aircraft Park.** |

the region encompassed by this book – Bircham Newton, Cranwell, Digby, Grantham, Manby, Mildenhall, North Coates Fitties, Peterborough, Sutton Bridge, Waddington, Wittering and Wyton, together with some large civil aerodromes, such as Sywell, allowing ordinary people a glimpse into the growing military aviation scene.

RAF Bircham Newton took pride of place for what might be considered the inaugural Empire Air Day on 24 May 1934, although because it was graced by a royal inspection visit, the station was not open to the general public. This date also fell on a Thursday, which was one of the reasons why attendance at the thirty-two participating stations was quite low. In subsequent years the air days were staged on the nearest Saturday to the actual Empire Day date.

While the national average gate number in 1934 was only just over 2,500, RAF Wittering for example, home at that time to the Central Flying School, attracted 2,000 visitors for whom the highlights of the show were formation aerobatic displays by Avro Tutors and a flying display by the new DH 86 four-engine airliner. Avro Tutors were also on offer at No.3 FTS Grantham's show, together with spirited aerobatics by Armstrong Whitworth Siskin IIIa single-seat fighters and demonstrations of message dropping and pick-ups by Atlas two-seat army co-operation aircraft. 1935 saw the same stations on the list but at Grantham the Siskins were replaced by an aerobatic display by Bristol Bulldogs.

A brief word about Empire Day may be of interest at this point. During the later part of Queen Victoria's reign it was felt that:

> Children should be reminded that they formed part of the British Empire and that they might think, with others in lands across the sea, what it meant to be sons and daughters of such a glorious Empire. The strength of the Empire depended on them and they must never forget it.

It was not until after her death that a special day was instituted in 1902 and the date of 24 May was chosen because it was Queen Victoria's birthday. Although celebrated thereafter across the empire, it was not until 1916 that it was officially recognised as an annual event. With the decline of the empire in the post-Second World War years political correctness took over until, in 1958, it was rebranded as British Commonwealth Day; in 1966 it became known as Commonwealth Day when the date was also changed to 10 June the official birthday of Queen Elizabeth II. In 1977 the date was changed yet again to the second Monday in March. So, Empire Day has officially ceased to exist for thirty years, except perhaps in the minds of those of a certain age and some Canadians who still keep 24 May as Victoria Day.

Providing light relief to both the participating stations and spectators alike, Empire Air Day air displays were the means by which the efforts of an expanding – and modernising – RAF were brought to the attention of the public. It was a way, too, for the Government to show what the taxpayer was getting for his/her money. Proceeds from entrance fees were donated to the RAF Benevolent Fund, a worthy cause, but was this not also just a thinly dressed-up ploy to ease the inevitability of war into the minds of the public?

Well, whatever the motive, on Saturday 23 May 1936, forty-nine stations threw open their gates for the public to enjoy themselves. For the first time since Empire Air Days began, No.3 Armament Training School Sutton Bridge, now with the status of a permanent RAF establishment, was included in the open day list. Almost 3,000 people visited the station, availing themselves of an opportunity to wander through hangars and even climb into aeroplane cockpits with pilots and ground crew on hand to answer questions. Despite the far from ideal weather – occasional showers and blustery winds, relieved by the sight of the sun from time to time – a varied and exciting programme was laid on.

Promptly at 2.15 p.m. three shiny silver Bristol Bulldogs in the colourful markings of No.56 Squadron took to the air, giving a spirited aerobatics display to open the show. Fresh from their success at Hendon, three equally colourful Hawker Fury aeroplanes of No.25 Squadron thrilled the crowd with precision formation aerobatics. In those days of relatively slow speeds, aeroplanes like these could be thrown around 'on a sixpence' and low cloud was not a problem although the blustery wind would tax the pilot's skill during aerobatics. Then came individual displays by various single Bulldogs, including one from No.56's CO, Sqn Ldr Leacock, who seemed to have lost none of his mastery of aerial manoeuvres learned on his way to becoming a First World War fighter ace.

Fighter attack techniques, the basis of everyday life at RAF Sutton Bridge, were ably demonstrated by a flight of Bulldogs, again from No.56 Squadron, upon a target drogue towed by the station flight's Fairey Gordon. A comic interlude of 'crazy flying' was staged with a DH Moth displaying 'L' plates. The stiff wind enhanced these hair-raising stunts but in the hands of Flt Lt McGill the little Moth appeared perfectly safe.

Naturally everything stopped for tea; the first part of the programme being brought to a close by a sedate formation of three station flight Gordons flown by Flt Lt Palmer and Sgts Drake and Roberts.

Air drill controlled from the ground by radio, followed by an ear-splitting dive-bombing routine by three Furies opening the second half. Somewhat more gently, a Hawker Hind was put through its paces by Sgt Roberts, while the final item on the flying agenda was a 'mass' formation flypast by nine aeroplanes; three each of Gordons, Furies and Bulldogs.

Meanwhile at ground level a Gloster Gauntlet was the focus of much interest as 'the latest RAF fighter'. Elsewhere the staccato rattle of machine gun fire tore through the air as the twin Vickers guns of a Fury, propped up in front of the firing-butts, were made to turn streams of lead into fountains of sand, much to the delight of the watching crowd. Armourers demonstrated how to fill ammunition belts, assemble and dismantle machine guns, while aerial photographers showed pictures of local landmarks taken from the air. The intricacies of wireless telegraphy were explained by the signals section and aeroplane servicing workshops opened for closer inspection. All in all there was much indeed for the visitors to see and enjoy that year.

Elsewhere in Lincolnshire the RAF put on similar air and ground displays at Waddington, where No.503 (County of Lincoln) Squadron, just re-designated as part of the Auxiliary Air Force, was based. RAF College Cranwell and RAF Digby, where No.2 Flying Training School operated, opened its doors to the public.

The event at Digby was somewhat overshadowed by an accident at the airfield on 21 May, the Thursday prior to the air show. One of the display items was to be a demonstration of the 'Flying Flea', G-AEBS, built by Sqn Ldr Charles Davidson MC, who died when the aeroplane crashed while he was practising for the show.

Empire Air Day at RAF Cranwell attracted a large crowd in May 1937. (Crown)

By contrast, the weather for the 1937 RAF Sutton Bridge Empire Air Day was ideal. Attracting 5,000 people this year the four-hour programme included even more variety of aeroplanes.

It was the turn of pupils from No.5 Flying Training School (RAF Sealand) to bring their Hart and Audax aeroplanes to the range for air firing practice in late May and early June 1937. They, too, supplied some of the thrills for the air show, held this year on 29 May. Pupils from that FTS opening the busy programme with a formation flypast of five Harts. All went without mishap today considering that during practice the previous day two Harts collided during landing, luckily each sustaining only minor damage and causing no crash.

Four Bristol Bulldogs flown by Flt Lt H. Eales in company with Sgts Smith, Wood and Stanford dived and wheeled repeatedly in simulated attacks upon a towed drogue target. Later, Sqn Ldr J.F.F. Paine displayed accurate low flying skill by dropping and picking up message bags using a hook attachment beneath his aircraft fuselage. Several solo aerobatic items were contributed by Flt Lts D.B.D. Field, N.H. Fresson, E.A. Springall and Plt Off J.D. Ronald; even the station commander, the irrepressable Wg Cdr Harry Smith, got in on the act with his own solo aerobatic effort. The inevitable comic turn was provided by Flt Lt V.Q. Blackden who made numerous flour-bomb attacks on an old car, driven erratically at high speed across the airfield.

By mid-1937 the Westland Wallace had replaced the ageing Fairey Gordon and a splendid formation of these new target tugs, lead by Flt Lt F.A.A. Strath, flew sedately over the airfield. A delightful show of syncro-aerobatics by Flt Lts Springall and J.C. Evans in a pair of Furies followed, before the day's proceedings closed with a grand formation flypast of three Wallaces, three Furies, one Boulton Paul Overstrand bomber from neighbouring RAF Bircham Newton and one Avro Anson.

Of all the Empire Air Days, those held in 1937 probably produced the finest of the classical type of show for the public and certainly those held in this region provided thrills galore. Despite its relative inaccessibility by road, RAF North Coates Fitties opened its gates to the public for the first time and attracted no less than 8,000 visitors, who were given a splendid programme that matched any of the other larger stations in its variety and content. RAF Peterborough's show drew over 4,000 while RAF Grantham hosted 3,000 visitors who, according to the *Stamford Mercury*, saw:

> A squadron of No 3 FTS planes cruising in 'pelican' formation; a Hawker Fury fighter demonstrating its manoeuvrability over a range of speeds before climbing to 4,000 feet and zooming into a power dive at 300 mph. This was followed by a flight of Furies making spectacular firing attacks on a sleeve target towed by a Fairey Gordon and the [flour] bombing of an armoured car.
>
> The air events began with a crazy-flying item, followed by a supply-drop of cargo – including a barrel of beer! – on to the airfield. A Hawker Audax then carried out message dropping and pick-up by hook at very low level. Five Hawker Harts excited the crowd with formation aerobatics while five more had great fun dive-bombing a Territorial contingent of the Lincolnshire Regiment in a mock battle. Other flying items were put on both by instructors and students in Avro Tutor and Hawker Tomtit trainers based at the station. The thrilling programme was brought to a close by the flypast of three Saro Cloud amphibian aircraft arriving from another airfield.

Not far away, Cranwell opened to the public for six hours, providing an attractive flying programme that included a last-minute addition of the brand new Bristol Blenheim flown over that afternoon from RAF Wyton.

A crowd of thousands, basking in brilliant sunshine at RAF Wyton itself, was greatly impressed by the sight and performance of the new Blenheims of 114 Squadron based at the station. Nine of these sleek monoplane bombers roared in towards the crowd, showing off their high-speed capability before breaking away in three 'vics'. Holding a tight formation, five of them then wheeled back and forth over the crowd as a prelude to the ever-popular 'guess-the-height' competition: a solo run by a Blenheim trundling over at 3,000ft.

In marked contrast two biplane Hawker Hind bombers of No.139 Squadron gave a sedate twenty-minute aerobatic display but undoubtedly the most graceful event was the formation aerobatic show by a flight of three Gloster Gauntlet fighters from No.80 Squadron.

The action-packed flying display was brought to a close by a grand flypast that included a Handley Page Heyford bomber, Avro Anson, Fairey Gordon, Gloster Gauntlet, Gloster Gladiator, Hawker Hind and a Bristol Blenheim – now what a sight that would have been!

However, amidst all the jollity, tragedy struck once again, this time at Waddington where the station flight had taken a Hawker Fury I, K8228, on charge only the day before the show it was scheduled to appear in the programme. Newly promoted Sqn Ldr Henry Power had moved from Waddington to Digby but, keen to help out, volunteered to do an individual aerobatics slot and flight-tested the aeroplane just before the show got under way. When his turn came he ran up the engine as usual, and then with everything seeming in order took off to begin his display.

Watched by a crowd of 5,000, the Fury climbed in an upward roll to 1,500ft when, according to his friend Flt Lt Eric Palmer:

> He began what I took to be a slow roll but instead of continuing this the nose dropped, which put the aircraft into a vertical dive. I cannot think of any reason for him getting into the dive. If he had been a novice it would have been more understandable but Sqn Ldr Power had more than 3,000 hours flying experience. He was only 600 or 700 feet up when the dive started and I thought he had no chance to pull out. It would be doing at least 180mph when it struck the ground.

Despite this blow the show carried on – now in front of a shocked and subdued audience.

The final air show at RAF Sutton Bridge was scheduled for Saturday 28 May 1938. In common with other aerodromes throughout the country, the display suffered continuous rain all day causing practically all flying items to be 'scrubbed'. Nevertheless a substantial crowd gathered, despite inches of water on the grass and tarmac areas, particularly braving the weather to inspect 'the new type all-metal bombers in the aircraft park.' These were probably Bristol Blenheims, some of which were based at nearby RAF Wyton at that time.

For RAF Sutton Bridge this was the end of an era in more ways than one. Its particular role was air weapons training and the station was gearing up for a war that was not far away. It was considered too important to have that impetus interrupted by something like an air show and so, after the 1938 washout, Sutton Bridge never opened its gates to the public again.

Just as we saw earlier, in the run up to the First World War, with all the newspapers now regularly carrying articles about civilian and military preparations for war, the public still turned out in their record thousands on Saturday 20 May 1939 to watch the RAF perform at Empire Air Day shows up and down the country. The programmes on offer differed little from the format of previous years but there were now more of what might be called the 'modern' types in evidence. RAF Hucknall sent a flight of Fairey Battle light bombers to the show at Waddington while, in addition to the Westland Wallace, Hawker Hind, Hawker Fury and Avro Anson, the new Handley Page Hampden bomber featured prominently in the station's programme. At RAFC Cranwell there were demonstrations of formation flying by its training squadrons' Airspeed Oxford and Hawker Hart aircraft but these were easily eclipsed by the appearance and performance of the Hawker Hurricanes of No.46 Squadron, based at Digby, while Waddington's Hampdens also put on a stirring show. The Hurricane and the new North American Harvard advanced trainer were also in evidence over at RAF Wittering.

At long last, then, it appeared that the RAF had some aircraft with which it could fight a war and as air historian Harald Penrose summed up the 1939 event:

Graceful aerobatics performed by three Gloster Gauntlets at RAF Wyton Empire Air Day in May 1937.

Britain was catching up. Even Munich seemed forgotten, despite dramatic evidence of massed military aircraft of every type displayed at the sixty-three Air Force stations on Empire Air Day. Air Minister, Sir Kingsley Wood with ACM Sir Cyril Newall flew a 465 mile tour of these stations [including RAF Waddington] in the Air Council's DH86B. It was obvious that monoplanes at last had won the day, though many biplanes were still in yeoman service. Every station reported crowds double that of the previous year and few drew less than 10,000. Including the fifteen civilian airfields involved, more than a million spectators were entertained that day.

Thus, what Sir Alan Cobham called the Heroic Age of Air Shows drew to a close and any aerobatics for the next six years would not have entertainment in mind!

CHAPTER SIX

Aces & Kings, Diamonds & Clubs

Edward, Prince of Wales, took the liveliest of interests in flying and he made every opportunity to fly in service and commercial aircraft to official and private engagements whenever possible. He received flying instruction during 1929 enabling him to pilot his own aircraft, and by 1930 had acquired a DH 60M and a DH 60X Gipsy Moth and a DH 80A Puss Moth G-ABBS. These three civilian-registered aircraft, the personal property of the prince, were hangared at RAF Hendon under the supervision of the prince's personal pilot, the RAF Reserve officer Flt Lt Edward Hedley 'Mouse' Fielden and ground engineer Mr Tom Jenkins, where, since the quantity was numerically equivalent to an RAF Flight, the unofficial term 'Royal Flight' began to be used when referring to them. Puss Moth, ABNN, was replaced by a DH 83 Fox Moth G-ACDD in 1932 but the following year the Prince was keen to own one of the new twin engine DH 84 Dragons and 'traded in' all three for G-ACGG, which was delivered on 12 June, luxuriously furnished and with an eye-catching paint scheme of red and blue fuselage, silver wings and red struts.

It was this strikingly coloured – and aptly named – Dragon, used for the very first time that day, which brought the Prince of Wales to the flying ground, in Exton's Road, King's Lynn, belonging to Sir Richard Bagge on Wednesday 21 June 1933 to open the annual Royal Norfolk Agricultural Show. A Puss Moth brought a bevy of press photographers from London and cinematograph film cameras whirred as the prince stepped from the aeroplane to be greeted by a veritable throng of local dignitaries, including the mayor of King's Lynn, and the legendary First World War fighter ace Wing Commander Raymond Collishaw, station commander of nearby RAF Bircham Newton.

Less than a year later Bircham Newton and Wg Cdr Collishaw once again played host to royalty when, on Thursday 24 May 1934, King George V and Queen Mary paid an official inspection visit to the station that also signified the inauguration of the first Empire Air Day. It was the first royal visit to an RAF station since the end of the First World War and the impressive flying display laid on included a variety of aircraft and a bombing demonstration by No.207 Squadron. On this occasion the station was not opened to the public, but the following year Bircham Newton joined in with the rest of the RAF by throwing open its gates to the paying public.

When King George V died at Sandringham House on 20 January 1936, Edward, now King Edward VIII, made another piece of flying history when he flew from Bircham Newton to Hendon the next day in his latest acquisition, DH 89 Dragon Rapide G-ADDD. He become not only the first British monarch to fly but the first to fly to his capital upon

Edward, Prince of Wales, meets local dignitaries upon arrival at King's Lynn in DH 84 Dragon, G-ACGG, on 21 June 1933 to open the Royal Norfolk Show. HRH is in a bowler hat seventh from right; First World War fighter ace Wg Cdr Raymond Collishaw is second from right; and the prince's pilot, 'Mouse' Fielden is fourth from right. (P.M. Goodchild via Ray Wilson)

accession. This Rapide was again in evidence when King Edward, accompanied by his brother the Duke of York – shortly to become King George VI – flew into RAF Wittering on 8 July that same year. The station and its resident unit, No.11 Flying Training School, had its turn to impress the new monarch on his first official inspection visit of his reign.

Being located alongside the A1 main arterial road would prove a mixed blessing for RAF Wittering, since its accessibility made it a popular place for other 'spit & polish' visits including some reciprocal visits by foreign delegations. One such visit took place on 29 October 1937 when a high ranking Luftwaffe delegation, comprising the German Secretary of State for Aviation, General der Flieger Erhard Milch, Chief of the General Staff, General-Major Hans-Jurgen Stumpff, and Director of the Technical Dept of the German Air Ministry, the charismatic First World War ace General-Major Ernst Udet – names indeed with a distinguished past and a notorious future – arrived at Wittering (plus Cranwell, Mildenhall and others) during a round of visits to several key aircraft, engine and RAF establishments.

It was the swastika emblem of the new Third Reich that was very much in evidence when twelve visitors in five German light aeroplanes flew in, on Saturday 12 May 1934, as guests of Sir Lindsay Everard, MP for Melton and chairman of the Royal Aeronautical Club Hospitality Committee, to his private airfield near Ratcliffe Hall. Ratcliffe aerodrome was home to a busy flying club alongside the Fosse Way, about eight miles north of Leicester, and when one of the German aeroplanes went missing it caused quite a stir in the corridors of power.

German pilots at Hendon. Heinkel Kadett and Focke Wulf Stieglitz aircraft assembled prior to departure for Ratcliffe airfield in May 1934. (Peter Green)

D-3224, one of the Heinkel He72 Kadetts that visited Ratcliffe airfield in May 1934. (Peter Green)

The German party first arrived in England at London's Heston Aerodrome then after lunch flew north to Ratcliffe, escorted by Everard's DH84 Dragon, G-ACEK *Leicestershire Vixen* in which two more German guests were carried, bringing the party up to twelve in number. They arrived with a flourish by buzzing the control tower 'and as they swooped to the ground the [red] swastika, the Nazi emblem, stood out boldly against a white background on the tails of the planes'.

The aircraft included two of the two-seat Heinkel He72 Kadets, one of which – registered D-3224 – was powered by a Siemens Sh14 radial-engine, and the other by an Argus AS8 in-line engine. In addition there were three of the two-seat Focke-Wulf FW44C Stieglitz, each powered by a Siemens Sh14 radial engine, two of which were registered D-2984 and D-2911. The next morning, wearing smart blue flying suits with a swastika badge, the visitors flew off to Mr Deterding's airfield near Daventry for lunch with Lord Willoughby de Broke, a keen private flyer, and it was during the return to Ratcliffe, taking in Sywell and Braunstone aerodromes en route, that one of their number became lost. But first, a little more background to the party itself.

It comprised Flieger Kommandanten von Bulow and Laumann, Flieger Kapitän Bieber, Flieger Schwarmführer Wegenast, Staatsrat Florian, Dr Weinhold, Dr Waldrich, Herr Bruegmann and Herr Scholz; female pilot Fräulein Margaret von Tresckow and Princess Matilde Windischgraetz, who went along as an interpreter. But the most important figure among the visitors was Flieger Kommodore Bruno Loerzer, president of the *Deutscher Luftsportverband* (DVL or German Air Sport Association).

Loerzer was already a force to be reckoned with in the 'New Germany' and his name would take on even more meaning in years to come – particularly where military aviation was concerned. Back in 1915, though, Bruno Loerzer was Hermann Goering's pilot while the latter was training to be an air observer in the German Air Service during the First World War. The two men formed a deep friendship that continued after the war when their fortunes in life were somewhat reversed. Loerzer had earned a worthy reputation as an air-fighting ace in the First World War, being credited with forty-one air victories and rising to command a fighter group. However, in those austere, post-war days in Germany, he became a cigar salesman while Goering, who turned out to be no mean pilot himself of course, cast in his lot with political firebrands like Adolf Hitler and the Nazi party.

The party's aims found much support among many former pilots and when Hitler came to power Hermann Goering began to formulate plans for a new German air force. In order to progress this he desperately needed an organiser and Loerzer's well-known personality and his proven organizational ability rekindled Goering's interest in his friend and he was recruited into the party machine. Loerzer's first task was to standardise civilian flying training procedure and bring the present profusion of flying and gliding clubs under a single central control called the German Air Sport Association, of which he became the first president. These were the days, of course, when Germany was not permitted to have a military flying organisation but it was Goering and Loerzer's aim to overcome this obstacle eventually, while publically trying to cultivate the opposite impression.

Interviewed upon his arrival at Ratcliffe, Bruno Loerzer said:

> The first enthusiasm of youth for flying is through the glider and thousands of them fly and construct their own gliders. Then they graduate to light aircraft. I have sixteen group centres of the Union and those attending are trained in aviation. Besides the teaching of aviation, comradeship is encouraged and everyone is taught to be unselfish, to learn to work for others instead of themselves. The whole object is purely educational and sporting. It is an endeavour to occupy the minds of the unemployed in Germany. There is nothing political or militaristic about it at all. For fourteen years this side of German life has been neglected and I am trying to do something to make up for lost time in this respect.

The use of militaristic titles and the capabilities of the aeroplanes on show caused *Flight* magazine to observe that '[the aeroplanes] possess more of the characteristics of high powered military aircraft', while *The Aeroplane* considered '[the DVL] covers the training of pilots on a scale with no parallel in Europe'.

German First World War ace and his party visit Leicester Aero Club in May 1934. Left to right: Lindsay Everard MP, Herr Bruno Loerzer, Herr Weginast and Flt Lt C.V. Phillips, in front of DH84 Dragon G-ACEK *Leicestershire Vixen II*. (*Leicester Evening Mail*)

Perhaps some of this training had been forgotten by Staatsrat Florian, because he took off on Sunday 13 May for the lunch trip with a companion on board but without a map. Moreover, it could have been the effects of a good lunch since, while heading back to Rafcliffe airfield on the leg from Daventry to Sywell, Herr Florian became separated from the gaggle of aeroplanes and lost his bearings quite badly. Instead of landing quickly, he just kept going until, with his petrol and oil almost exhausted, he did eventually do the sensible thing and landed to find out where he was. It was 6 p.m. by the time he came down in a seed field on Captain Wilson's farm in Surfleet Marsh, about three miles north of Spalding but fifty miles north-east of Sywell and fifty miles east of Ratcliffe! Having no knowledge of the countryside and only a little English he tramped across the fields to a main road, found a house with a telephone, established his location and telephoned Ratcliffe Airfield to let them know what had happened and to organise a rescue. His demise had caused quite a flap and aeroplane search parties from Ratcliffe and Desford airfields had been buzzing all over the Midlands for several hours trying to find the missing pair without success.

Was this just an innocent case of getting lost or was it a good excuse to have a look for military installations for future reference and maybe take a few photographs? There was no reason at all why Florian could not have landed much sooner than he did; to get from the Northampton area to Spalding he would not have had to wander far off track

'Get lost!' Two German pilots and a Heinkel He72 Kadett at Ratcliffe Airfield in May 1934.
(*Leicester Chronicle*)

to have easily taken in, for example, RAF Wittering and RAF Peterborough. Whatever the motive, on Monday 14 May 1934 the visitors returned, via Heston and dinner at the House of Commons, to Germany.

Six years later, Blitzkrieg in France began. 'On May 13 1940 General Guderian's panzers and infantry crossed the river Meuse ... General Bruno Loerzer's Fliegerkorps II gave the army continuous air support to great effect.' Loerzer and Goering et al had achieved their objective.

In the aftermath of the Second World War, Spalding Airways began life as one man's dream of how air travel for pleasure would grow as the effects of austerity lessened in Britain. The driving force behind this air taxi and aero club venture was George Clifton, a Lancashire man who set up an electrical business in south Lincolnshire in the mid-1930s. He learned to fly as a private pilot in 1936 and on the outbreak of war joined the RAF. Having completed sixty-two operational bomber sorties and with postings as a staff pilot with the Empire Air Armament School and a Maintenance Unit test pilot, by the time the war ended George Clifton had considerable experience of flying aeroplanes in all conditions.

On 22 May 1947 George collected his airline's first aeroplane from Tollerton Aerodrome (Nottingham) and flew it to a small grass field owned by Mr T. Mawby at Wykeham on the outskirts of Spalding. It was a three-seat Auster J-2 Autocrat registered G-AJIU — a far cry from the mighty Halifax bombers he used to fly. Ted Wellband carried out aircraft servicing for him in a small shed in one corner of the field. Later on, when the business became more settled, a larger field further down Spalding Marsh was leased, where a blister hangar and an office were erected and where it was hoped to create a flying club offering flying lessons.

The much travelled Auster J/1 Autocrat G-AJIU, first aircraft operated by Spalding Airways.
(Ted Crampton)

One of the first excursions made by the Auster was when George and Ted took Mr J. Baker, an enthusiastic private pilot who lived in Gosberton, to 'The Products Of Normandy' air rally organised by L'Aero Club de Caen et du Calvados at Caen/Carpiquet Airfield in France on 14–15 June. The flight took just over six hours in poor weather, which at times forced them down to 100ft over the Channel. But it was all worthwhile as they won a prize for being closest to their pre-declared arrival time out of the twenty-seven competitors.

Nevertheless, it was in the fields of business flying and private charter work that Spalding Airways grew most rapidly and with George's long-range bomber experience, no trip was considered too far. The Auster's range on full standard tanks was about 400 miles and there were plenty of airfields available in those days. Early flying jobs were a variety of short-range, ad hoc passenger air-taxi jobs and small cargo trips, often on behalf of the farming community. South Lincolnshire is famous for its bulb growing industry and in March 1948 a local businessman took full advantage of air travel to corner a part of the bulb market.

A new variety of daffodil had been successfully bred in Cornwall and this exciting news came to the attention of one of Lincolnshire's leading growers who recognised a good business opportunity when he saw it. He wanted to see the new flower immediately and chartered Spalding Airways' Auster to fly him down to Cornwall. So impressed was he with the daffodil – called *Lizard Light* – that he bought the entire stock of 200 bulbs on the spot and brought them back to Spalding in the aeroplane.

Later that same year, 1948, Spalding Airways undertook its longest charter. This involved a number of flights to Italy by new pilot Jim Crampton, another local ex-RAF bomber

Spalding Airways pilot Jim Crampton at the controls of Auster Autocrat G–AJDZ. (Ted Crampton)

Spalding Airways' Auster Autocrat, G–AJDZ, after a mishap at Spalding Marsh airfield in 1949.
(F. Bollons via D. Benfield)

pilot. Jim not only took the little Auster *India Uniform* to Florence and back, a round trip of about 2,000 miles, but also made the round trip twice in ten days. The reason for the flights was that a Mr and Mrs George Bland of Spalding were going to Italy on holiday and had hired Spalding Airways to transport them out and back. The Blands were an ordinary couple, of modest means, but they possessed an overwhelming desire to travel and lived quite frugally in order to indulge their wanderlust.

Fully laden with passengers and luggage, the Auster took off from Spalding at 8 a.m. on a Monday reaching Florence, by way of Dijon and Nice, early on Wednesday morning – a leisurely pace but one that allowed time for a little sightseeing en route! Jim Crampton then left Florence at 1 p.m. that same Wednesday with one passenger, Mrs Toniolo, the English wife of an Italian businessman living in Spalding. This return leg was completed by 6.30 p.m. the next day with one overnight stop en route. The following Thursday Jim went back to Florence, this time to collect Mr and Mrs Bland, with whom he departed on the Friday, and arrived back at Spalding airfield the next day. Quite a marathon for pilot, passengers and aeroplane!

The flexibility of the little airline made it popular with local businessmen and it soon became necessary to move a short distance to a bigger, sixty-acre field on Crowtree Farm, to allow slightly larger aircraft than the Auster to be used. At various times during the next three years, six different Austers (G-AJIU; AHHP; AIPU; AJDZ; AJYB; AJYM) were operated by Spalding Airways in addition to such classic aircraft as the Fairchild Argus 2 (G-AJPC), Miles M38 Messenger 2A (G-AJFF) and twin-engine Miles M65 Gemini 1A (G-AKDK) and de Havilland 89 Dragon Rapide (G-AEMH). When the company moved from Spalding to Westwood Aerodrome in Peterborough and changed its name to Peterborough & Spalding Airways, it also operated a DH 89A Dragon Rapide (G-ALBH, formerly an RAF Dominie), DH 104 Dove 2 (G-AOYD) and Piper PA-27 Aztec 250 (G-APXN; ARBR). Owners of these aircraft included many well-known enterprises such as Baker Perkins, Mitchell Engineering, Mitchell Construction and Geest Industries.

Surely an early example of that modern concept, the 'weekend break', is the first trip made in the newly-acquired DH Rapide, G-AEMH. This came at the end of April 1949 when Jim Crampton flew a party of eight – Mr and Mrs Barker and Mr and Mrs Wray of Donington, Nurse Fox, Miss Arden, Mr Ruysen and Mr Ruygrok of Spalding – from Wykeham Airfield to Holland for a weekend. Landing in The Hague after a flight – routed via Lympe – of just under four hours, the party enjoyed a tour of the Dutch tulip fields by car, visiting flower festivals in Lisse and Hillegom, then some of the group explored the canals of Amsterdam by boat while others went to a seaside resort. Leaving The Hague at 4 p.m. on the Sunday afternoon, in less than four hours they were back in Spalding again. Perhaps it was the success of this sort of trip that prompted George Clifton to found Spalding Travel Agency, a business that prospered on the rising tide of holiday travel over the next two decades.

9–10 September 1949 saw Spalding Airways fly its busiest weekend period since it was founded. In forty-eight hours, pilots Clifton, Crampton and Clark flew three aircraft on 14,000 passenger miles. Passenger flights left Spalding airfield for Nottingham, Jersey, Guernsey and Dublin in addition to several freight flights, including one to Manchester to collect vital spare machinery for a combine harvester belonging to south Lincolnshire

G-AEMH, DH Dragon Rapide of Spalding Airways landing at Spalding Marsh airfield in March 1947. (Ted Crampton)

SPALDING PLANE IN NORMANDY RALLY

A Fairchild, piloted by Mr. George Clifton, Spalding Airways, was tenth in the third international air rally organised by the Aero Club of Basse-Normandie at the week-end. Our photograph was taken at Spalding airfield before some of the members took off on Saturday. Left to right : Mrs. C. R. Shinner, Mr. Shinner, Mrs. E. W. Dryden, Mr. Clifton, Mr. Dryden, Master Ken Dryden. Another Spalding Airways aircraft, a Rapide, piloted by Captain N. W. Gray, also went. The passengers were Mr. G. Neal, Mr. L. Massey and Miss J. M. Harwood.

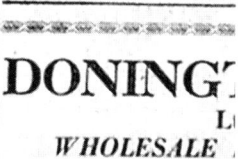

DONING'
L
WHOLESALE

A weekend in Normandy by Spalding Airways. Pilot George Clifton and well-known local travellers preparing for take-off from Spalding Marsh airfield, *c.*1948. (*Lincolnshire Free Press*)

farmer J.W.E. 'Bill' Banks. The previous week five Auster and Messenger aircraft from Northants Aero Club made a social visit to Spalding Airfield. In common with many in the south Lincolnshire area, the visiting pilots were all farmers, showing just how popular flying had become among that fraternity.

By 1950 George Clifton considered that to be successful in providing air passenger services you either had to remain a very small operation or become a very large one with heavy financial investment and borrowing. Neither option appealed to him, though, so very astutely he changed the direction of his flying business. Clifton broke new ground by becoming one of the first British companies after the Second World War to recognise the potential of executive business flying. In conjunction with Peterborough-based Mitchell Engineering Company and Baker Perkins Ltd, already regular clients of Spalding Airways, George Clifton leased the former RAF airfield at Westwood, Peterborough, from the Air Ministry and built up Spalding Airways into a company specialising in business aviation management. The first landings at Westwood airfield were made on 16 December 1950 and flying operations from Spalding Airfield were gradually scaled down. The last commercial landing at the Spalding Marsh Airfield, before operations were completely transferred to Peterborough, took place on 10 July 1951 when George Clifton flew the Miles Messenger in from Liverpool.

The business blossomed and during 1951, with two pilots, Clifton and Norman Gray, flights were made carrying passengers to most European countries, in addition to over 500 tons of freight and covering 100,000 revenue-paying miles, a large part of it for Mitchell Engineering. Aircraft servicing was still carried out at the Spalding field by Chief Engineer W. Proctor.

In 1960, in order to reflect the expanding operation, Spalding Airways was re-formed into Peterborough and Spalding Aviation Ltd (P&SAL) with four equal partners – George Clifton (who remained MD), Mitchell Engineering Ltd, Mitchell Construction Ltd and Baker Perkins Ltd. It owned no aircraft of its own but was responsible for the operation of those owned by the business shareholders – at that time two Aztecs and a Dove. By 1962 two new pilots had been recruited and the pilots that were, at various times in the 1950s and 1960s, employed by the company were Cedric Platts and Norman Gray, both later captains with BEA; Alan Watson, later an air traffic controller at Heathrow; P.A. Patrick, later captain with International Air Charter Co.; Duncan Macintosh, later captain and MD of Loganair; and Jim Crampton, later MD of Norfolk Airways and Air Anglia.

In 1965 P&SAL moved its flying operations from Westwood – which would soon disappear under housing development – to the former USAAF bomber base at Glatton just south of Peterborough where, in December 1966 due to a general downturn, the business was finally wound up. Mitchell Engineering continued to operate one Aztec, G-AWDI, at Glatton and business flying there developed under the banner of Peterborough's Business Airfield with Klingair coming to prominence in more recent times. George Clifton's travel agency in Broad Street, Spalding, carried on for many more years while George himself went to live on the Isle of Man and took long winter breaks at his apartment in Palm Beach, Florida.

Interestingly, Clifton's original little Auster, *India Uniform*, did not just slip quietly off into some scrapyard in the sky. It is unclear quite what it it got up to in the intervening years but it caught the eye of this writer at an RAF Waddington air show where enquiries revealed that in 1999 Martin Greenhalgh, a private pilot from Sheffield, bought Auster G-AJIU from someone in Doncaster. It was in a poor condition but Mr Greenhalgh painstakingly restored it to flying standard and it was given a certificate of airworthiness that year. Now he flies *India Uniform* himself and is a frequent visitor at airshows where he proudly displays the little Auster restored to its full glory.

Another of George's Austers had its moment of (in)glory when it became forever linked with the opening of Boston's public airfield in 1950. As mentioned earlier Auster J-5B Autocar, G-AJYM, was originally bought by Mitchell Engineering Co. Ltd in January 1950 and operated on its behalf by Spalding Airways. Although he did not explain why, according to George Clifton the Autocar was found unsuitable for Spalding Airways' requirements and was returned to Auster's Rearsby factory where it was used as a demonstrator – at least until its demise in April of that year.

'BOSTON IS NOW ON AIR MAP' was the press headline following an inspection of the Boardsides field on Thursday 6 April 1950 by officials of the Ministry of Civil Aviation. Boasting two grass landing strips, each of 500 yards length, Boston Airfield was pronounced fit to be licenced as a public airfield, opening the way for fifty years of flying operations at the site. The airfield, located in an area known as Wyberton Fen, came into existence in late 1948 as a base for Boston Air Transport Ltd (BATL) which, during the next year, provided aircraft for the training needs of the embryonic Boston Aero Club in addition to air charter work and pleasure flying, which continued until 1954 when that company's flying operations ceased. BATL is known to have had Miles Messenger 2A, G-AJDM and G-AKKK registered to it around this time.

Ill-fated Auster J/5B Autocar, G-AJYM, pictured at Boston airfield just prior to its final flight in 1950. (Peter Green)

After a winter break, flying training restarted in the spring of 1950 with Peter Towell going solo and Les Hewitt becoming the first member to qualify for his private pilot licence (PPL). By the date of the first fly-in the club had 180 members, twelve students had gone solo and four had qualified to PPL standard. The granting of the public licence was duly celebrated when the club was officially opened on Good Friday, 4 April 1950 by the mayor, Cllr E.C. Stanwell, at a grand 'At Home' fly-in event.

Over 1,000 spectators watched the arrival of forty aircraft attracted in from airfields such as Derby, Leicester, Nottingham, Peterborough, March, Wolverhampton, Elstree and nearby Skegness and Spalding. First down at 11 a.m. was a green Auster from Derby, followed shortly after by two aircraft from Spalding flown by Ted Wellband and George Clifton. A blue, twin-engine DH Rapide (G-AKOG or AKMD) piloted by Skegness Aerodrome manager Peter Bushby landed next, soon to be joined by the Skegness Fairchild Argus, G-AJGW. By 1 p.m. nineteen aeroplanes, including Austers, Messengers and Geminis were lined up on the airfield.

In his speech the mayor announced that the newly formed Boston Air Transport Company would take delivery of a twin-engine aircraft, possibly a Miles Gemini or a DH Dragon Rapide, for commercial operations. Then, wishing the Aero Club 'the best of luck and good, safe flying', he cut a tape across the clubhouse entrance to mark the official opening. True to the long-established custom at such events, the mayor, his wife and club president Richard Hardy were then taken for a short flight in a Miles Messenger piloted by the CFI Mr G.N. Snarey. Greatly excited by what was her first flight the mayoress, addressing the crowd over the loudspeaker system, said 'I am sure it was thrilling. It has

DH Dragon Rapide, believed to be G-AKOG, operated by Skegness Air Taxi Co. Ltd from Boston and Skegness *c.*1950. (Max Hundleby)

Skegness airfield manager Peter Bushby with Fairchild Argus II, G-AJGW, at Skegness airfield, in the Butlins colour scheme and with the town's 'Jolly Fisherman' emblem on the fin. (Max Hundleby)

been wonderful. No one need be nervous about flying, especially when you have got such a good pilot.' Tragically, both those optimistic mayoral sentiments were cast to the wind with the death of all three occupants of Auster Autocar, G-AJYM, that crashed outside the airfield perimeter just one hour later.

Piloted by Stanley Bradshaw, a journalist and artist on the staff of *Aeroplane* magazine, accompanied by Edwin Riding and Norman Stoneham, both of whom held 'A' licences, the Auster had landed just before midday after a flight from Elstree and was making its departure at 4 p.m. when the accident happened. The four seat Autocar was the latest model produced by Auster Aircraft Ltd at its Rearsby factory and AJYM, said to be the prototype of the model, was borrowed from the makers the previous day. Mr Bradshaw, an experienced pilot who learned to fly in 1926 and served with the ATA during the Second World War, had flown it from Rearsby to Elstree, then flew his party up to Boston on Good Friday morning where his attendence was as correspondent for his magazine.

The crowd watched the Autocar take off, make a circuit and then fly slowly across the airfield at about 700ft towards the adjacent river and railway line. A *Boston Standard* reporter recalled what he saw next:

Suddenly the nose went up in the air, the plane half turned and slipped into a spin, diving to earth. There was a tense split second, which seemed like an hour, before the helpless machine hit the ground in a cloud of dust. The blue tail appeared to rock forward then fell back onto the river bank on the opposite side to the railway, 300 to 400 yards from the airfield. The crowd was shocked into silence. A fire engine, ambulance and doctor's car rushed over Wyberton Chain Bridge and drove headlong across ploughed fields to reach the scene. The plane did not catch fire but it was a complete wreck with one wing torn off that lay shattered at the foot of the bank. The doctor pronounced all three occupants dead at the scene.

Boston Airfield also became home to Aerial Spraying Contractors Ltd in 1950, with its fleet of four Austers configured for aerial crop spraying operations. That company was absorbed by Skegness Air Taxi Service Ltd which also took over the running of the airfield upon the demise of BATL, in addition to its operation at Skegness. Lincs Aerial Spraying Co., another subsidiary of the Skegness business, took over responsibility for crop spraying aircraft and general aircraft maintenance facilities in a move that set the pattern of operations at Boston for the remainder of its time as an airfield.

Most weekends the novelty of the airfield usually drew a small gathering of spectators eagerly waiting to see what aerial visitors would turn up. On Sunday 23 April 1950 patience was rewarded, for example, by the arrival of a red Piper Cub from Huntingdon, an Auster Autocrat from Denham and Peter Bushby bringing eight passengers in a Rapide on a trip from Skegness. Most interesting of all, though, was the arrival of Mr L.R. Bulling from Great Hucklow in his light blue and maroon Olympic 1 glider named *Peveril*. Circling the airfield and losing height rapidly, the glider took many by surprise as it 'swished' over the lane and clubhouse to make a gentle landing at the far side of the field, earning the honour of being the first glider to land on Boston airfield.

Many willing hands rushed to help the pilot push the glider back to the clubhouse, where Mr Bulling rang Great Hucklow to report his safe landing and order up a retrieval trailer. Then there was time for a cuppa and to regale the onlookers with details of his flight:

> I was in cloud for about ten minutes and had no idea where I was until on breaking cloud I spotted RAF Cranwell five thousand feet below, then I decided to make for Boston. I was just going to drop into a field on the north side of the town when I saw the name 'Boston' in huge letters on the roof of the Air Transport Company's hangar. Seeing the parked aircraft I realised I had reached the aero club and so I came down here. I travelled about 73 miles.

1951 was the year of the Festival of Britain and Boston was just one of the towns and cities throughout the land celebrating a feeling of release from post-war austerity. The aero club organised a grand Festival Air Display for Saturday 16 June, announcing a programme of 'aerobatics in abundance, parachute jumping, aerial demonstrations and pleasure flights from 10am to 6pm'. The *Boston Standard* reported 'BOSTON AIRSHOW WAS TOPS. A Thrilling Festival Spectacle – With One Hitch.' The 'hitch' referred to was a delayed take-off for the mayor and mayoress, Cllr and Mrs J.P. Roe for their celebratory, free ride. After the opening ceremony the two VIPs boarded an all-black, twin engine Miles Gemini where they settled comfortably for pilot Wilf Pearson to whisk them aloft. But the starboard engine refused to self-start. Airfield manager Cliff Anniss intervened and swung the prop several times but still there was no spark of life from the engine. Up stepped Cllr E.A. Moffatt to the mark. He gave the prop 'a heftier swing' but no joy. However, Cllr Moffat was a force to be reckoned with and so, in the end, brute strength did the job and the engine coughed into life. Now the port engine came out in sympathy and it, too, refused to start. Cllr Moffatt by now had the bit between his teeth and gave that prop an almighty heave – and the engine fired but even with both engines running, the Gemini was still not ready to roll as the wheels had embedded in the grass and it needed a good shove to set it rolling. At last they were off, followed into the air by a yellow

Miles Messenger carrying the Festival Queen, Mary Thomas, and her two attendants, Jean Miller and Barbara Read.

The air programme opened with that old favourite of air shows, a height and speed judging competition flown here by Bill Holderness, with his daughter in the back seat of his blue Miles Magister. He soon had the spectators agog when he made a fast dive 'to what seemed like inches from the ground. The rumour that daisies were afterwards found in his engine was denied by an official spokesman!' Even Bill could not match the speed of the next item, though. Although an RAF Meteor jet was scheduled in the programme, a DH Vampire suddenly appeared above the airfield ten minutes early. To the relief of the organisers the Vampire kept out of the way and did not hang around and the same applied to a glider, spotted hovering near the airfield. Right on cue, the Meteor woke up the crowd when it zipped across the field – very low and upside down! The pilot pulled the jet up into a near vertical climb, rolled off the top into a dive at the field, pulling out at the last minute into a climbing roll then a loop, finishing off with a hestitation roll in front of the spellbound crowd. Off he went leaving the stage empty for Auster's chief test pilot, Ranald Porteous, to show off Auster J-1 Aiglet, G-AJYW.

Built with the agricultural market in mind, it was highly appropriate that the Aiglet was shown off so well at Boston, because of the eighty-six built only fourteen stayed in the UK and of these, seven were owned by Aerial Spraying Contractors Ltd of Boston. Each winter from late 1950 until 1954 these Aiglets were ferried out to the Sudan for anti-mosquito spraying operations by British pilots, one of whom was Spalding Airways' Jim Crampton, who spent several winters out there. The Aiglets were subsequently acquired by Skegness Air Taxis Ltd but continued to operate from Boston in an agricultural role.

Auster J/1B Aiglet, G-AJYR, while in service as a crop sprayer with Aerial Spraying Contractors Ltd based at Boston. It was written off in an accident in July 1964. (Peter Green)

In the late 1960s the name 'Spalding Airways' came into brief existence again but it had no connection with George Clifton's earlier business of that name. This time it was more of a one-man-flying-by-the-seat-of-the-pants outfit, operating from a grass field adjacent to J.W.E. Banks' St Guthlac's Lodge farm in the district known as Postland (Crowland), and it is most likely that the occasional air charter simply helped with the upkeep of the aircraft rather than anything more sophisticated.

The aeroplane involved was a single-engine Miles Messenger, G-AKIR, a comfortable four-seat monoplane that was registered to J.W.E. Banks & Sons Ltd in October 1968 and appears to have remained in use at its Crowland field until 1971. 'Chief pilot' at that time was Ray Fixter who was quite a character, with a reputation for his somewhat flamboyant approach to the art of flying. This writer recalls a flight in the Messenger with Ray at the controls, during which we had an interestingly-close up view of Holbeach parish church weathercock! It is also known that AKIR was used on several occasions to convey local businessman John van Geest senior and his wife, when he needed to visit his Atlantic shipping fleet base at Barry in south Wales. The destination on those trips was Cardiff (Rhoose) Airport which is just a few miles from the port and while working in Geest accounts department, this writer recalls handling invoices headed 'Spalding Airways' for those trips.

When gliding operations began at Crowland Airfield, Ray Fixter also used to fly the resident bright yellow Taylorcraft Auster 6A Tugmaster, G-ARIH, from the passenger seat of which this writer has also been able to count the individual leaves on the poplar trees that used to grace the Queen's Bank section of the circuit! ARIH was just about able to tow a single-seat glider into the air and carry two people without stalling. It was unusual

Miles M38 Messenger 2A, G-AKIR, was based at Postland airfield during the 1960s.

in being one of only three of this type that had two side by side seats and full dual controls for pilot and passenger. Auster 6A Tugmasters were civil conversions of the army AOP Mk6 model but all other conversions retained the space alongside the pilot that previously housed the army radio gear and the radio operator/passenger seat was in the back of the cockpit, at right angles to the fuselage side.

Although Ray faded away from the local air scene he did not fade away entirely because, rather curiously, he ended his days as curator of the National Cycle Museum near Grantham.

In the UK, the sport of gliding had gained in public popularity during the 1930s and from time to time these relatively rare 'birds' were seen in the wide open spaces around The Wash, where they were something of a novelty. In fact, post-1945, purely private flying in this region in general was very sparse for many years, although tucked in amongst the region's intensive military air activity there was an occasional civilian flying incident that caught the eye of the local newspapers. For example, there is evidence of a farm airstrip – the most likely sort of venue for private flying in those days – being in existence at Farcet near Peterborough in the late 1940s. It was noted that on 7 June 1950 a Miles Messenger 2A, G-AHZU, with two occupants had a mishap when visiting Mr K. Whittome at his farm. The aeroplane landed heavily on the farm airstrip, wrecking its undercarriage and damaging one wing in the process.

Gliding, though, has remained popular in the region with clubs springing up and gliders 'running out of air' and dropping in on farmers from time to time. Great Hucklow in Derbyshire and Dunstable Downs in Bedfordshire have long been the two principal Meccas for gliding enthusiasts in the UK since before the Second World War. In fact, way back on 19 April 1936 one of what must have been Great Hucklow's first members, Mr A.L. Slater, landed his single seat glider at Gosberton Clough, near Spalding, having flown from the Derbyshire airfield.

It took until Friday 14 August 1953, however, before a pilot from that airfield provided another glimpse into the state of gliding in the UK at that time. The glider was first noticed by J.W. Driver of Wisbech who was out delivering newspapers in the Saturday Bridge area. After watching it for a while, Mr Driver continued with his round but the glider always seemed to keep up with his van until, at last, it circled and landed in Tom Biggadike's field at Holbeach Fen, where its sudden arrival frightened a pack of horses quietly munching in one corner of the field. Safely down, the pilot stepped from the cockpit, bid Mr Driver a cheery 'Hello' and announced that he was Mr C.R. Adams and he had flown from Great Hucklow, a distance of eighty-seven miles.

It was during a competition at Great Hucklow, home of the Derbyshire & Lancashire Gliding Club since 1935, that Mr Adams had been winch-launched off its famous cliff. Encountering excellent soaring conditions he decided to try his luck to see how far he could fly and it was only when he ran out of up currents near The Wash that he had no option but to land where he did. While waiting for his recovery team there was plenty of time for his glider, *Speedwell,* to become a source of curiosity for the local population.

One year later, Great Hucklow was the venue for the prestigious world gliding championships and in 1954, this gave rise to the fens seeing a rare slice of the action. This competition, which comprised of a number of individual and team races in various

HAPPY LANDING IN A FIELD OF HORSES
Mr. C. R. Adams stands by his glider

Mr C.R. Adams poses next to his glider 'Speedwell' in August 1953. (*Lincolnshire Free Press*)

categories, dates back to 1937 when it was first held in Germany; 1954 was the first time it came to the UK. The only other year it has been held in the UK was in 1965 when RAF South Cerney in Gloucestershire was the venue.

On Wednesday 21 July 1954 many heads turned skywards at the sight of a veritable flock of gliders circling silently near The Wash. Their pilots, competing in the world championships had set off from Camphill Airfield at Great Hucklow and were now looking for a safe place to land, no doubt before they drifted towards the sea. Coloured bright red, the first, flown by Phillip Wills, a member of the British team, landed safely in a field near Langwith Drive, Holbeach. At about the same time a glider flown by one of the German team, Herr August Wiethucher, narrowly avoided colliding with high tension electric cables as he came down in a field belonging to Mr R. Tinsley at Holbeach St John's. Other gliders dropped in around Wisbech and one landed at Terrington while, near March, Swedish team member Per Persson was making his safe return to earth and was in fact later declared the winner of the event, having flown a distance of ninety-seven miles.

Phillip Wills is one of the few UK glider pilots to have been awarded the prestigious Lilienthal Gliding Medal and the Paul Tissandier Diploma, by the FAI in 1954 for services to gliding and sporting aviation.

By the 1960s Fenland Gliding Club at RAF Marham was one of many clubs flourishing in the region. In July 1960 the *Lynn News & Advertiser* carried a story:

It is confirmed that Flt Lt G J Rondel, a member and instructor at the Fenland Gliding Club at RAF Marham, has become holder of three gliding records. On June 18, in the club-owned Olympus IIB glider, Flt Lt Rondel was towed up from Marham and after being released over Swaffham, flew over The Wash on a routine cumulo-nimbus flight. During the flight he broke three records – the British national gain of height record of 28,200 feet; the UK local gain of height record of 27,600 feet and the UK absolute altitude record of 28,500 feet. Although Flt Lt Rondel thought he had equalled the world gain of height record of 30,100 feet this latter was not ratified.

Another snippet dated 6 May 1966 showed that Fenland Gliding Club was still going strong when pilot Mr Wilton-Jones, attempting a 50km flight, had to drop in to a wheat field at Sutton St Edmund after making 42km. He did, however, find time to have tea with Peter Jones of Luttongate Road while he waited for his retrieval team.

It seems that gliders dropping in for tea were becoming quite the norm. On 20 May, for example, Plt Off. Peter Maclachlan was attending a week's gliding course at No.644 Air Cadet gliding school at RAF Spitalgate, near Grantham. Gliding conditions were favourable so he settled himself comfortably in the cockpit as he soared up to 4,000ft off the top of a winch launch. His aim was to fly to the disused former RAF airfield at Sutton Bridge, a distance that would give him another BGA qualification. With his goal almost in sight he ran out of lift and, down to 1,000 ft, he needed somewhere to land – and quickly. As far as the eye could see there were growing crops but he did spot one field that seemed to be free of vegetation. There were power cables down one side but it had no hedge near a road on the other side so that would be handy for the recovery crew to get the glider out. He brought the Sedburgh glider down safely and that is how he came to be taking tea in Pinchbeck smallholder Charles Lacey's garden.

It was back in 1968 that the sport of gliding first took root at Crowland and one of the early members, Mike Stillingfleet, kindly made available his account of the development of gliding at the airfield. He was keen to emphasise the vital role played by the late J.W.E. Banks, who Mike considered to have been a good friend to the club and without whose generosity and encouragement it would never have survived.

In March 1968 Perkins Gliding Club took up residence at Crowland airfield. Their previous home was the former wartime USAAF airfield at Spanhoe (Station 493) between Stamford and Corby. Spanhoe Airfield, however, had come 'under the hammer' the previous winter when the concrete runways were broken up for hardcore. This was just the first phase of development at the airfield since, by 1980, the western side of the airfield was being eaten away in the quest for iron ore to feed the steel works in Corby.

Preliminary enquiries by club secretary Jack Lovell paved the way for club officials to negotiate with Mr Banks – a keen aviator himself – for the use of the existing airfield which was home to Mr Banks's aircraft and the small Spalding Flying Club. With his encouragement and help the club decided to undertake a trial period at Crowland. One reason for the hesitancy was because it was thought that the site might prove unsuitable due to being in such a flat fenland area only fifteen miles from The Wash. However, after the first season, despite worse than normal weather conditions, it was decided to accept Mr Banks's offer to use the airfield until at least 1970 and then review the situation again.

During the winter of 1968–69 the hangar used at Spanhoe Airfield was dismantled and re-erected at Crowland Airfield. At the beginning the gliders were winch-launched and the field was shared with a flock of sheep, which at least kept the grass in reasonable shape. Late in 1968 aero-tow facilities were made available thanks to the efforts of Derek Wilcox and Harry Feneley using either a Tiger Moth or Beagle Terrier, which was ferried over from Cranfield Airfield for weekend flying. In order to make gliding a more viable proposition at Crowland it was felt that more permanent aero-tow facilities were desirable. Harry Feneley therefore decided to buy the Beagle Terrier from his other syndicate partners and based the aircraft at Crowland. The fact that he lived at Little Staughton near Sandy in Bedfordshire did not seem to deter him.

The club fleet at that time consisted of four gliders: a Slingsby T21B open cockpit, two-seat training glider; an Eon Baby, open cockpit, single-seater for early solo pilots; and two single-seaters, an Eon Olympia 2 and a Slingsby Skylark 2.

With aero-tow facilities now available on a regular basis, the feeling was that a more modern two-seat glider was needed. A five-man syndicate, headed by club stalwarts Reg Bradshaw and Hayden Haresign, bought a new Bocian IE glider and it arrived at the airfield on 16 May 1969. The situation was further improved when Phil Cracknell, an ex-member of the club, decided to base his Skylark 4 at Crowland and, as he did very little gliding himself, generously permitted suitably qualified pilots to fly it. This aircraft arrived in June 1969 and the following month was joined by an Italian Morelli M100S single-seater.

Just at the point when the 'fleet' was increasing nicely, disaster struck. Two accidents occurred on the weekend of 12–13 July that resulted in the Slingsby T21B (BGA765) and the Eon Baby (BGA628) being severely damaged to such an extent that they were treated as 'written off.' Fortunately no injuries were sustained by either pilot. On a brighter note, on 27 July the first two recorded five-hour Silver C Leg flights were achieved. Hayden Haresign flew the Bocian to complete his Silver C and Harry Feneley did his in the M100S; he also cloud flew to 8,000ft during that flight.

Thus, by the end of 1969, gliding was firmly established at Crowland Airfield with many soaring flights achieved; the myth, that the fenland area has a scarcity of the thermal activity needed to extend soaring flights, was well and truly dispelled.

Serious problems, though, loomed on the club front. Perkins Gliding Club (PGC) was down to one single-seat glider, the Olympia 2, and the Skylark 2 was actually owned by Baker Perkins Sports Club (BPSC) – a different entity – whose members flew as associate members of PGC. PGC was negotiating for a Bocian two-seat glider to replace the two aircraft written off; however, at this juncture PGC also wished to restrict the numbers of 'non-PGC' members to get a more equal quantity of each. As the current ratio was about two to one in favour of BPSC and the majority of gliders were non-PCG, this caused an awkward situation. It was perhaps understandable that Perkins Sports Association – which existed and was created for the benefit of its members who were employees of Perkins Diesel Engine Company of Peterborough – should not wish to appear to be financing a venture where its own members were outnumbered!

Discussions at committee and club level produced no agreement and, following a special meeting, a decision was taken reluctantly by the PGC camp to form a separate, open club to be known as the Peterborough and Spalding Gliding Club. The British Gliding Association, whilst sympathising with those members wishing to form a new club, had to support the existing club which was a BGA member and since the BGA was opposed to a new open club being formed and operating from the same site, a further stalemate was reached. In the meantime Perkins Gliding Club had taken delivery of its new Bocian, which arrived in May 1970.

The objective of the proposed new, unrestricted club was to further the sport of gliding in the area and to open up membership to the general public. That the club actually 'got off the ground' on 1 June 1970 was due in no small measure to the approval, enthusiasm and financial generosity of Bill Banks. It did, after all, operate alongside the existing Perkins Gliding Club under the control of Roy Taylor, PGA chief flying instructor. Founder members included: Tony Fidler, chairman; Jack Lovell, secretary; Gordon Figg, treasurer; Reg Bradshaw, chief gliding instructor; Hayden Haresign, deputy CFI; and Harry Feneley, tug pilot.

The P&SGC started with virtually no assets, since the Bocian two-seater was owned by Baker Perkins Sports Association, the Beagle Husky was owned by Harry Feneley and a syndicate owned the M100S. Initially there were twenty-five members, of whom only three were qualified instructors – Reg Bradshaw, Hayden Harsign and Tony Fidler.

Another setback occurred on 26 September 1970 when the Bocian was severely damaged in a landing accident at the end of an instructional flight. The glider was badly damaged but, once again, fortunately no injury was sustained by the occupants. Perkins Gliding Club kindly stepped in and offered to handle P&SGC trainees in its aircraft while the Bocian was sent away for repair.

The club desperately needed a hangar and once more Bill Banks came up trumps. He just happened to have a dismantled hangar at his other airstrip at Witham-on-the-Hill and offered it with no strings attached, providing club members collected and erected it themselves, which they were only too keen to do. That hangar is still in use.

By the end of the first year the new club membership had risen to fifty but were operating with only three instructors. To increase that number a BGA Instructor course was run at Crowland by the new assistant national coach, John Heath, and five suitably experienced pilots were persuaded to undertake training.

During May 1971 John Bowles and Mike Stillingfleet took the Perkins Bocian over to Norfolk for the glider flying week held annually at Swanton Morley Airfield. It turned out to be a glorious few days. On 18 May, launching at 10.30 a.m. and landing back at 7.18 p.m., Alf Warminger, flying his Phoebus 17, became the first to complete a UK 500km out and return from Swanton Morley. This spurred Tony Fox and Tony Fidler into taking the M100S in a trailer to Swanton the next morning to get a piece of the action. The final day, 21 May, saw Tony Fidler flying in the M100S, in company with John Bowles and Mike Stillingfleet in their Bocian back from Swanton to Crowland, both gliders arriving safely but not without an anxious moment or two en route. To cap a memorable week, Tony Fox completed a five-hour flight in Norwich Soaring Group's Skylark 2 to qualify for his Silver C duration award.

Things were beginning to take shape back at Crowland, too. The hangar was re-erected by the end of February 1971 and a Tarren concrete sectional building, purchased for use as a clubhouse, was erected alongside it, with a generator installed in the hangar to provide electrics for both buildings. Next the club negotiated a grant from the Sports Council and in August bought the Bocian from the syndicate. The syndicate in turn purchased Phil Cracknell's Skylark 4 which, with the M100S, became the second aircraft to be owned by a group comprising Reg Bradshaw, Hayden Haresign, and Tony and Gordon Figg. Another syndicate consisting of Jack Wayman, Tony Noble and Fred Logins purchased a Slingsby Capstan, a side-by-side two seat-glider which they made available for instruction and general club use. In August 1971 Perkins Sports suffered a mishap when Olympia 2B, BGA618, was badly damaged after it broke away from a winch tow at about 75ft and landed heavily, fortunately without hurting the pilot.

In order to promote gliding in the local area, local newspapers were invited to send reporters to the airfield and the club had some useful publicity following a visit by one of the local radio station's roving reporters. Added to this, there were a number of 'passenger evenings' arranged for the employees of various organisations to sample a gliding experience.

More aircraft were added to the 'fleet' in 1972 with the arrival of a Slingsby Skylark 3, purchased from the Husbands Bosworth Club by a syndicate and a Slingsby Tutor owned by Mike Stillingfleet. Realising the vulnerability of relying on only one tug aeroplane, Harry Feneley bought Auster Tugmaster, G-ARIH, which became known as the 'Yellow Peril' for obvious reasons. Although 1972 saw a lot of changes of aircraft ownership it was all a strong indication of the club's growing vitality. Fred Logins bought out the Capstan from Jack Wayman and Tony Noble who themselves acquired the M100S from Tony Fidler and Brian Essam; the latter then bought a renovated Schleicher K6E from Dunstable Gliding Club. A second K6E was purchased by a new syndicate comprising Norman Brown, Eric Goodwin, Tony Noble and John Delahoy. Thus, by the middle of 1972 the Spalding and Peterborough Gliding Club had grown to operate two tug aircraft and owned a Bocian, two thirds of a Capstan and a Skylark 2. There were also six syndicate gliders, namely an M100S, Skylark 3, Skylark 4, two Schleicher K6E and a Tutor, operating under the club banner.

As for the airfield, despite initial doubts, it proved to be a reasonably good thermal site with soaring flights too numerous to mention except to say they resulted in many Silver C certificates over the years. Several 300km triangles were flown as early as 1973-74 with members Fidler, Haresign and Goodwin being the first to obtain Gold C distance and the coveted Diamond Goal certificates.

Over the intervening years, up to the present day, the club has gone from strength to strength with a membership of about ninety. There are two club-owned tug aircraft, three two-seat gliders for training and early solo flights and two solo gliders. In addition, at the last count, there were twenty-six syndicated or privately owned gliders. The club has held an annual open day since the 1990s and is keen to promote gliding to young and old alike. A thriving cadet scheme is overseen by mentor, instructor and part-time tug pilot Gerry Pybus who, along with his wife Joan, typifies the many hard-working people who have striven to make the Peterborough and Spalding Gliding Club a success.

It would be remiss not to mention some of the other gliding clubs that have contributed so much to flying for fun in the region, such as the Nene Valley Gliding Club, which has since 1982 conducted its operations from a field that was once the site of RAF Upwood runways.

The 'Yellow Peril.' Auster 6A Tugmaster, G-ARIH, at Postland airfield.

Competitors in a gliding competition seen at Postland airfield in May 1981.

Straddling the Lincolnshire/Leicestershire county boundary is Buckminster Gliding Club, which operates from another former wartime base, the old US transport airfield of Saltby between Grantham and Melton Mowbray. Similarly, Lincolnshire Gliding Club is based on the former RAF airfield of Strubby, near Alford, and of course there has been a gliding club on the grassy acres of the north airfield at RAF College Cranwell since the 1950s. Further afield, RAF Syerston (Newark) and RAF Spitalgate (Grantham) were home to RAF gliding schools for many years.

For a short time in 1979 a 600-yard stretch of grass field on Roads Farm, Kirton, belonging to Mr Charles Bowser, was used by a gliding syndicate but this seems to have discontinued after one of the members, Reginald Knight from Burgh le Marsh near Skegness, died in a launching accident there on 14 July that year.

Quite often, so it is said, real life mirrors fiction and in some ways the birth of a new airfield for private flying in the middle of the fens might be thought of in that way. For a cracking good thriller yarn with an aeronautical flavour, track down a novel written back in 1966 by ex-jockey turned best-selling author Dick Francis, called *Flying Finish*. Coincidentally, the storyline features a 'Fenland Airfield' located near The Wash in south Lincolnshire with a resident 'Fenland Flying Club.' All this was, in fact, some six years away.

Reality had to wait until September 1971 when a group of real-life local aviation enthusiasts founded their own airfield and club where they could learn to fly, and despite the proximity of Postland and Sibson airfields, they wanted something different. Rejecting alternative names for the new enterprise, like the 'Meridian Flying Club' (nearly sitting on the Greenwich meridian), they settled on the 'Fenland Aero Club'.

Local farmers were contacted to ascertain which of them would be willing to make some land available and eventually Mr Bernard Wright agreed to an airfield on his land at Whaplode Lodge, Holbeach St Johns. It was arranged that his son John, together with a Mr John Moore, who already had a pilot's licence and lived locally, would construct the grass runways and that the new club would pay them rent. That winter, members of the new club set to and built the first timber club house, only to find that, come March, a gale blew it down. Undeterred, it was soon replaced by an asbestos-clad building that soldiered on for many years. Meanwhile, the spring of 1972 saw the runways being seeded and by September that year the airfield including the demanding (i.e. short!) 26/08 runway, had been licensed by the Civil Aviation Authority (CAA). Indeed, in those days, the telephone wires still ran above ground on the approach to runway 26, making it even more tricky than it is today. Dick Yeates prevailed upon RAF officer Brian Willcocks, then instructing at RAF Hemswell, to become the club's first chief flying instructor (CFI).

A year after that first meeting, the club held its first open day, to which the public could come and see what was on offer. On 11 November 1972 training started in earnest in Paul Coulten's Jodel D120A, G-ASXU, at £6.75 per hour. Without dual brakes, the Jodel was always a test of the best instructor and the following May it was abandoned in favour of the Rollason/Druine Condor D62B, G-AWSR, rented from Michael and Norman Jones of Redhill and the first of several Condors to be resident at the field. Petrol cost 38p a gallon in those days and flying training was only available at weekends, but even this received a setback when Brian was posted to Germany. Nevertheless, club founder member and secretary Fraser McEwan became the first person to obtain his pilot's licence at Holbeach St Johns airfield. Training was able to continue when Assistant Flying Instructor 'Sam' Salmon joined the club, helped by some judicious cover from Sibson's CFI, the well-known and highly experienced Barry Tempest.

Complaints from neighbours began to crop up around this time; for example, one from Whaplode St Catherine wrote to East Elloe Rural District Council complaining about

Brian Willcocks, first CFI at Fenland Aero Club, with Jodel D120 trainer G-ASXU, February 1975.
(Fenland Aero Club)

'the noise from the aircraft constantly over flying [his] home' to the north of the airfield.
Fortunately the objections were not upheld. It was at this point that local Councillor
Geoff Hare, who had always voiced strong support for the club, was persuaded to become
a member and remains one of its most stalwart characters.

The year 1973 brought problems in the form of an international oil crisis and in November
that year, when its effects began to bite, flying was banned on Sundays. Fuel deliveries were
generally uncertain at this time, but fortunately some members had a secret store in drums,
squirreled away somewhere in Holbeach St Johns. Training was given a boost in May 1974
when Sam obtained his full instructor's rating and that year saw Bob Whyte, John Scott, Mo
Fowler and Geoff Hare obtain their PPLs from examiner Dave Hughes.

Winter operation was always uncertain at this time since no drainage was installed for
the runways but, in fairness, the original group of would-be aviators (Dick Yeates, Dave
Buffham, Fraser McEwan, Cis Perrin, Geoff Southwell, Dave Ladbrooke and Fred Hamlin)
could hardly have foreseen the ever-increasing utilisation of the airfield. It became quite
common, therefore, to see water standing on the main runway from November onwards
so decisive action was taken and the first drainage system was installed.

Sunday 8 September 1974 was a landmark that saw the very first Fenland Aero Club
air show take place and even high winds and squally showers failed to put a damper on
its success. This was also the first air show to be seen in the South Holland district since
those heady days of Cobham et al during the 1930s, and like those far-off days, it attracted
several thousand people.

A spirited aerobatics display by local farmer Lindsey Walton in his Nord 1002 Pingouin/ Bf108, G-ATBG, masquerading as a German Messerschmidt Bf109 gave many thrills, especially to those who were old enough to remember the Second World War. The strong wind caused a number of items to be cancelled, including a Spitfire, hot air balloon, parachute drops and a glider display by members of the nearby Postland club. Flour bombing and crazy flying by a biplane caused much amusement but the highlight of the afternoon was the full, competition-quality aerobatic display by Philip Meeson in his Pitts Special. Among the other pilots taking part in the show were Barry Tempest, leader of the Barnstormers Flying Circus, and Sqn Ldr Wally Epton. The aircraft included a Tiger Moth, Stampe, Stearman and Tipsy Nipper. Though modest in scope, this first show was judged to be a great start for what was intended to become an annual fund-raising event

In February 1975 Brian Willcox returned to the UK and took up his position as CFI once more. Meanwhile, powered flying training finally ceased over at Postland (Crowland) airfield with the demise of the Spalding Flying Club; their stalwarts, such as Ray Bettinson, joined Fenland Aero Club although, as has been mentioned elsewhere, gliding activity still carried on at Postland. In the new mood of confidence, Brian Willcocks decided to take the plunge and in June he set up the 'Fenland Flight Centre'. Until then the club had operated the aircraft, with the instructors performing on a voluntary basis, but from now on the club agreed to sub-let the flying training and supply of fuel, in exchange for membership fees from each student.

June 1975 also saw the first 'Spamcan' on the airfield when original student then Postland convert John Parker returned to Holbeach St Johns with Cessna 150, G-AVNC. A second 150 arrived in December when club landlord, Richard Wright, who had gained his own PPL at Sibson, made Aerobat G-AYRO available to the centre. Flying training was now available at Holbeach St Johns five days a week. Despite Brian's allegiance to tail-draggers, the economics of actually running a business spelt the end of the faithful Condor. It was not really suited to operation from a grass field and the brakes issue caused continual problems. The end came when Cambridge club member, Doug Hopper, agreed to provide Cessna 150s, G-BAZR and G-BBXB, on a fulltime basis in spring 1976.

'An air show in the old tradition of death-defying aerobatics, crazy flying and more sober pleasure flights', ran a report of the 1975 Fenland air show in the *Spalding Guardian*. Opening at noon on Sunday 21 September and graced with better weather than the previous year, the entrance fee of just £1 for a car and all its occupants was terrific value for money and 4,000 people took advantage of the opportunity. John Day and his team worked hard to put together an entertaining aerial programme; it was all hands to the pumps to help handle the spectators and cars on the ground, while Ray Bettinson organised another team looking after catering and concessions.

Fresh from its success in the Concours d'Elégance competition at a recent European light aircraft rally was an immaculate DH Hornet Moth, G-ADKK, totally rebuilt by managing director Cliff Anniss in the workshops of the Lincs Aerial Spraying Co. at the Boardsides Airfield in Boston. Its stately display could be contrasted with Barry Tempest's usual hair-raising flour bombing and crazy flying routine, and when Lindsay Walton put his Nord 1002, G-ATBG, through its paces he used all his skill and every inch of the airfield at very low level, to show off its manoeuvrability.

DH Hornet Moth G-ADKK, owned and restored by Cliff Anniss, manager of Boston airfield, was a frequent attraction at Fenland Aero Club's air shows in the 1970s and early 1980s. (Tony Hancock)

Flying training accelerated when club member Bob Lyons obtained his assistant instructor's rating and joined the staff of the Flight Centre, while Don Cousins set up in business as resident engineer in the north east corner of the hangar. Hitherto, fuel had been dispensed from a 250-gallon bowser but this was replaced by a 1,500-gallon underground tank. Further improvements were made out on the airfield with lighting and Litas units being purchased from Shipdam Airfield and installed down the runway. Another milestone arrived when it was decided to drop the long-winded title of 'Holbeach St Johns Airfield' and, quite against the normal practice used for naming airfields after the nearest settlement, the CAA permitted the change to 'Fenland Airfield'. The ICAO code 'EGCL' was allocated and the new runway lights became licensed shortly after that.

1976 saw the airfield expand to accommodate training seven days a week, when Bob Lyons obtained his full instructor rating. With Brian Willcocks remaining as CFI, there was now the luxury of having two fully qualified flying instructors as basic training switched from Condors to Cessna 150s. The change in trainer type did not really come as a surprise when one considers that only the year before it had taken no less than three and a half months to complete the C of A on Condor, G-AYFF, and a staggering ten months to complete that of G-ARSW.

The hourly training rate on a Cessna was now £14 and for a Condor it was £13. There was also a Piper Cherokee 180, Bob Whyte's G-ATNB, available for training. It was still a sad day though when, on 19 July 1976, the last Condor tail-dragger, G-AYFF, left the airfield. Meanwhile, at the airfield pump the price of petrol rose to 90p a gallon.

In glorious sunshine, 5,000 spectators craned their necks to catch the first glimpse of red smoke marking the descent of the private freefall parachute display team called the Kestrels. The four-man team, consisting of former RAF Falcons member Mick Geelan, Chris Thompson and ex-'Red Beret' instructor Tony Keogan, was led by another ex-RAF Falcons member Joe France, and made a thrilling start to the 1976 Fenland Air Show held on 5 September that year.

In addition to all the regular displays by vintage and modern biplanes, the club committee really excelled itself this year by persuading a number of famous acts to perform for the show or at least to route over the airfield for one or more spectacular passes. Following the Kestrels came a series of solo aerobatics and a flour bombing routine by the irrepressible Barry Tempest in his Tiger Moth; this was followed by Ernie Cunningham, an instructor from the Norfolk Gliding Club at Tibenham, in a Skylark 4 glider and Dave Parfrey in the ex-Air Cadbury Pitts Special, performing for the first time at an air show. After a height judging competition there was more parachuting by the Kestrels. Sutton Bridge farmer and expert display pilot Lindsey Walton played Second World War in his Nord 1002/Me108, authentically painted in Luftwaffe markings, and a Piper Pawnee 235 crop sprayer from Lincs Aerial Spraying Co. at Boston showed just how low these chaps really flew over the local fields. Barry Tempest was back in action showing off the aerobatic qualities of a Japanese Fuji FA200-160 trainer flown in to Fenland by Geoff Rennoldson, and stepping out of one cockpit and into another his next slot enabled him to put the Northair (Leeds) Cessna 150 Aerobat through its paces.

Within ten seconds of their predicted time, the wide, open countryside echoed to the deep roar of the Lancaster, Spitfire and Hurricane of the RAF Battle of Britain Memorial Flight, which swept over at low altitude several times. The Lancaster was in the steady hands of the latest BBMF commanding officer and the RAF's Hastings expert, Sqn Ldr Ken 'Jacko' Jackson. When the roar of the Merlins had faded into the distance, Boston's Cliff Anniss, in his immaculate DH Hornet Moth, G-ADKK, brought things down to a more leisurely pace, until all eyes strained skywards once more to pick out the four gaudily painted Pitts Special biplanes of the Rothmans aerobatics team, who brought the show to a close with a flourish.

Unfortunately the winter of 1976–77 turned out to be exceptionally wet and to all intents and purposes the airfield was effectively shut down from November to March. With Doug Hopper now running the flying school and Bob Lyons as his CFI, training managed to limp along with the kind co-operation of the McCaully Flying Group who allowed the club to operate out of Little Snoring Airfield where they had the advantage of the hard wartime runway.

Meanwhile John Wright, following the damage done by hurricane winds on 2 January 1976, was busy repairing the hangars to enable Don Cousins to open for business in June 1977. Neighbour trouble flared up again when one actually drove a car onto the runway to make a strong point. This happened just as John Day was taxying out, causing police and the CAA to investigate the incident but the rather lurid accusations made against the club were found to be unsubstantiated.

Around this time, too, the club went through an unpleasant period of accidents. It began innocuously enough with damage to the nose wheel of G-BBXB on 3 July 1977.

Much worse though was to come, on 6 April 1978, when one of the most popular student pilots, Malcolm Robinson, met with a fatal accident in G-BAZR. In tribute, John Day presented a trophy named after Malcolm, which is competed for each year on the club's Day of Events. 15 December 1978 also saw the loss of founder member Fraser McEwan while he was attempting an ILS approach into Luton airport in zero/zero weather. In 1979 G-BBXB was again damaged when the prop hit a block left on the airfield, while 1978-arrivals G-ATMM and G-AWAX were both involved in incidents. AWAX ended up in the drain at the end of runway 26 and was written off, fortunately without personal injury. In fact, though, AWAX rose like a phoenix a few years later when someone had the bright idea to repair it and convert it to a tail-dragger.

On a lighter note, on Sunday 4 September 1977, the committee really excelled itself by laying on a vintage air show that surpassed even the previous year's magnificent event. Over 7,000 spectators just about filled the airfield to capacity and they thrilled to the breath-taking sight of the RAF's world-famous Red Arrows aerobatic team making a seventeen-minute display over this small grass airfield. Nine bright red Hawker Siddeley Gnats under the leadership of Sqn Ldr Frank Hoare filled the sky with dazzling manoeuvres and lots of red, white and blue smoke, at times seeming to be just feet above the heads of the enthralled spectators. With no obstructions for miles around the Red Arrows certainly pulled no punches when their routine called for low passes!

The RAF was particularly well represented at this show since Fenland proved to be on a convenient route for several aircraft transiting between bases and other shows that year, and the committee had gone to enormous efforts to pre-arrange fly-bys to fit into the programme. Possibly the noisiest of these was a Nimrod maritime reconnaissance

Cessna 152 G-AWAX after its last flight with a tricycle undercarriage! (Fenland Aero Club)

aircraft, en route from Cyprus to its base at RAF Lossiemouth. The open fields allowed the mighty Nimrod to delight the crowd with a couple of very low passes, one with the bomb doors gaping open.

The closest RAF airfield to Fenland is Wittering, home of the Harrier V/STOL fighter. Always – even nowadays – seen at low level, all across the fens, one of these remarkable fighters didn't disappoint as it screamed in from Wittering at about 500mph across the airfield, showing off its agility before slowing down for its party piece of a vertical landing followed by a vertical take off, from which it rose ever higher until at 8,000ft it accelerated away towards Wittering.

A little too much cloud and wind caused the cancellation of the hot air balloon, Luton Minor and RAF Falcons parachute items, but by way of compensation the Falcon's Hercules transport aeroplane, which had been circling patiently for forty-five minutes at 5,000ft over the airfield waiting for an improvement in the weather, dropped down to the 'deck'. Making a low, slow pass down the main runway with the big rear cargo door wide open, the Falcons team could be seen inside, waving like mad as it went by.

Having battled through some pretty rough weather over the Pennines from another display, the ever-reliable Lindsey Walton arrived on time to put his Nord 1002/Bf108, in the colours of the Luftwaffe's Molders Geschwader, through its usual high-quality aerobatic routine. Then it was the RAF's turn again, with the growl of Rolls-Royce Merlins heralding the arrival of the BBMF Lancaster, once again with Sqn Ldr 'Jacko' Jackson in command, flanked by a Hurricane and a Spitfire. As the BBMF completed its sweep around the airfield, another deep sound, familiar to those in the area who could remember the daylight bomber formations of the Second World War, grew into the silver shape of the Boeing B-17G Flying Fortress *Sally B*. This was an unexpected addition to the programme and pilot Don Bullock – a man noted for his low-flying exploits – caused many to wonder if the committee had found a novel way to cut the airfield grass! So low were some of the passes that even the highly experienced Barry Tempest, who himself flew in three of the display slots that day, was moved to observe 'Don Bullock was flying *Sally B* and on one pass from south to north picked up some straw either with his wingtip or the tip vortex. Absolutely horrific! Shook me rigid at the time.' Photographers who were quick on the draw got some remarkable pictures of the B-17 that day! Barry's own contribution to the show was in using all his own experience to display the aerobatic capabilities of the Rogers Aviation Cessna FRA150 Aerobat, G-BCFR, and the Fuji FA200, G-BBRE, and to present his highly controlled but very entertaining 'crazy flying' and toilet roll cutting act in the brightly coloured Tiger Moth, G-ANZU. Philip Meeson put the Pitts Special, sponsored by Smitty Perfume, through its paces and then a Rumanian IS28 glider, in the capable hands of Ernie Cunningham, completed the aerobatics on offer. Bizarrely, this same glider was reported to have spent the previous Sunday under 60ft of water at the bottom of a gravel pit! Members of Norfolk Gliding Club salvaged it after the crash and spent all week working against the clock to dry it out and repair it in time for the Fenland show. Towed aloft by a tug aeroplane from the nearby Peterborough and Spalding Gliding Club at Postland airfield it showed no sign of its earlier mishap. The very varied programme was rounded off with good displays by a locally based Enstrom helicopter, a Pawnee crop sprayer from Lincs Aerial Spraying and

How low can you get? B-17 Sally B in the memorable, ultra-low-level beat-up of Fenland Aero Club at its air show on 4 September 1977. (Via Ray Nicholson)

a Percival Prentice, one of only a few of the former RAF 1950s trainers still flying in the UK, flown in from Scotland for the show by Bob Batt.

The show was acknowledged by organisers and spectators to have been a great success and later, by some observers, to be possibly the best ever seen at Fenland. This is probably the closest one could get to emulating those vintage air shows of the 1930s. With the entrance fee held at the very attractive sum of £1 a car it still made a handsome profit for the club which, when added to the coffers, went a long way to financing a number of improvements to the airfield facilities in subsequent years. As a result of the healthy state of those coffers, spring 1979 saw the installation of the second phase of the runway drainage system and further plans were already under discussion.

For some time members had been grappling with the possibility of a homing device for Fenland in order to overcome the difficulties, for both members and visitors, of finding the airfield! After several requests, National Air Traffic Services (NATS) visited the airfield during a rather fortuitous snowstorm and agreed to allocate the airfield a Non-Directional Beacon (NDB) frequency. The uncharacteristic speed of this decision making might have had something to do with the possibility at that time of a British Airways helicopter visiting the airfield every night on a GPO mail run, or perhaps with the decision of the RAF to give up its own beacon at RAF Barkston Heath. Whatever the reason, fortune smiled on the club when Fenland was granted 395kHz and with the kind assistance of John Wright, a beacon was installed in his scrapyard adjacent to the airfield. By 1990 this NDB frequency had changed to 401kHz and is still (in 2007) going strong. It took a week for a JCB to dig the trenches for the thirty-six radial copper wires, essential for the beacon ground mat. British Airways, no doubt with its own interests uppermost in mind, agreed to loan the club a transmitter, which was duly collected by air from their airfield at Beccles. The 'FranAir MH50' set was soon installed on the back seat of a derelict Hillman Imp car at the base of the 50ft aerial mast, the mains electricity connected – and a car window left open for cooling! Then, on 1 December 1979, the UK's latest navigational aid came on air and at a stroke the number of calls from lost pilots was drastically reduced. Despite the potential contract between BA and the GPO never actually materialising, the NDB remained in place and Fenland went on the map as probably the smallest airfield in England to have its own navigational aid.

Yet more improvements were in the pipeline. The only parking area for aircraft, up to now, was located between the signal square and the road. Recognising the need, the landlords agreed to rent the club an acre of land (half a hectare) to the south of the signal square and this was levelled, seeded and brought into use in time for the 1980 Strawberry fly-in. Profits from the series of air shows during the 1970s enabled the committee to replace the old timber toilets, control tower and storeroom with the present splendid goldfish-bowl tower, new kitchen and toilets. The room beneath the tower was named the McEwan Room in memory of the club founder who had died in an air accident. At the same time and following the loss of their office in the old clubhouse, Doug Hopper bought and erected the wooden 'Flight Centre' building next door, which gave faithful service until it, too, was replaced in 2006.

Apart from the 'main event' air shows each year there were two other regular events organised by the club. These were the Flower fly-in, which coincided with the famous Spalding Flower Parade at the beginning of May and the Strawberry fly-in, held each July to coincide with the harvesting of that local delicacy of the fenlands around the airfield. These events are still running in the twenty-first century (2007), and not only is the airfield open to all aerial comers, but the public also has a great opportunity to view a wide range of visiting light aircraft arriving from far and wide – with a prize for the pilot that flew in from the greatest distance.

Lunch was provided for those arriving by air for the Flower Parade, after which they were gathered up and taken to and from the parade route in Spalding by coach. All arrivals by air for the Strawberry fly-in in July received strawberries and cream and any pilot touching down on the enormous strawberry painted on the runway won double helpings for all on board! In 1978 no less than seventy aircraft, including a twin-engine Dornier DO28D Skyservant, flew in for the July event in glorious hot sunshine. The 1979 event was equally blessed with fine weather although the number of aircraft fell to fifty. Weather conditions were obviously the great decider on how many visitors would drop in and on several occasions the visiting aircraft were between fifty and 100, but the record seems to have been 1986, when the crews of a grand total of 109 aircraft consumed a prodigious amount of strawberries and cream.

The 1978 show, held on 3 September, had to go a long way to match the previous year but John Day and his committee certainly made a valiant effort to do so, and still all for a £1 entry fee. With the ever-popular RAF Falcons parachute team opening the show, the core of the programme followed the tried and tested formula of previous years.

New items were always difficult to find but the committee kept coming up trumps with interesting turns, such as Rob Lamplough's Harvard, a type that was a regular sight in the same piece of sky during the 1940s and '50s, and the big American training biplane Stearman PT19, G-AROY, owned and flown by Johnny Jordan. New to that year's event were the Royal Navy Historic Flight's Fairey Firefly Mk4, WB271, and Hawker Sea Fury, TF956. The former was the only example in the UK but was sadly lost in a fatal accident at Duxford in recent times, while the latter is one of the fastest piston engine fighters ever made and with the arrival of a BBMF Spitfire for its display, the public could easily compare the attributes of each.

Undoubtedly the star of that year's show was Flt Lt David Cyster who flew in with his Tiger Moth, G-ANRF. David had not long returned from his solo flight in the Tiger Moth from England to Australia, made in June 1978 to commemorate the first such flight by the legendary Bert Hinkler fifty years before. The spirit of Alan Cobham was still alive, too, when David gave a short flying demonstration during which he made a steady pass over the runway for the 'Guess the Height' competition, which was won by Alison Smith of Spalding with her guess of 365ft. Her prize was a flight in one of the aero club Cessna aircraft. It sounds rather blasé now to say that the show was brought to a close by the usual splendid Red Arrows display.

Spring 1979 saw the arrival of good weather that prompted members to turn their thoughts towards longer trips and with this in mind the club arranged an aerial outing to its twin club at the St Rémy Airfield near Sézanne, France. Seven pilots set out and all arrived safely, although the bare patches on the wing-leading edges bore testimony to the bad weather encountered en route. In the years to come there were many reciprocal visits between the two clubs, continuing even to the present day (2007).

The new goldfish bowl control tower was commissioned over the winter 1979–80 and gave those in command a 360° panoramic view for miles over the surrounding countryside. All of this would have been ideal for the annual air show on 2 September, but it was not constructed and lowered into position until towards the end of the year.

The show was again blessed by a spell of good weather which enabled John Wright to organise the critical task of baling the straw in time to clear the area used for car parking. A new act in the shape of the Army Red Devils parachute display team opened proceedings this year. They were followed by an eclectic mix ranging from a twin-engine Partenavia air taxi; Bristol Plane Preservation Unit's record-breaking Miles Gemini, G-AKKB, flown by James Buckingham; the 'Katabatic' Pitts Special in the expert aerobatic hands of Ian Senior; a Harvard and a Hawker Sea Fury flown by Wallace Cubitt and Spencer Flack respectively. In stark contrast to these lively items was the Volmer Jensen VJ-23 rigid hang-glider flown by David Cook. His was no fly-by-the-seat-of-the-pants display though, since over the preceding five years David had earned a considerable reputation as an exponent of hang-gliding but was also recognised as a pioneer in the field of powered hang-gliding; he was the first pilot to fly such a craft across the English Channel, in 1978.

After the popular interval 'Guess the Height' competition there were more biplane aerobatics from Stearman, G-AROY, Tiger Moth and Stampe, then newcomers in the form of an Auster AOP9, all followed by the Tiger Club's Druine D31 Turbulent formation display team from Redhill Airfield, whose lively low-level antics included a unique synchronised crossover manoeuvre through a hand-held hoop just 14ft high!

Barry Tempest, local pilot Lindsey Walton and Fenland's own Bob Lyons each had a share in the three-hour, action-packed programme that ended with another heart-stopping display by the B-17 Flying Fortress *Sally B*. Dry spring weather was ideal for the Flower fly-in and the new grass aircraft parking area was fit for the Strawberry & Cream event, although it would not yet stand up to use in wet conditions. Later in the summer weather prospects looked good for the seventh annual air show scheduled for 31 August 1980. However, in the two days running up to the show the rains came and it stayed wet until the day before, leaving the whole event in doubt because no straw baling could be done on the car park field.

The morning of the show dawned fine and dry with a slight breeze. Fred Campling swung into action with the baler and just managed to keep one step ahead of traffic arriving from late morning for the show. The RAF was noticeable by its absence for the first time in four years. The programme explained:

> With the price of aviation fuel rocketing to £2.00 a gallon, the price of the visiting acts has naturally increased and we have been compelled to raise the admission price accordingly to £1.50 but we trust that you will feel that you have had value for money when it's time to leave. We have combed the land to bring you as many acts as our budget will permit. Unfortunately the national mood of restraint and cutback has meant the absence of any RAF participation this year but we have several military aircraft for you nonetheless.

'Just add water' was the almost appropriate description of the opening item – the Pernod Sky Diving Team, comprising four men and four women parachutists dropping in from a Britten Norman Islander aircraft. Once again the show committee had worked hard to alternate 'regular' display items with some new blood. Designed in the late 1960s, the only example of a specialist aerobatic biplane, known as the Cranfield A1, was shown off by one of the UK's leading display pilots, Eric Steenson. This unique aeroplane could be contrasted with the similar Pitts S2a Specials of the Rothman team whose members, Marcus Edwards, Rod Rea, John McLean and Brian Lecomber, gave solo as well as team formation displays. Neil Jansen brought his 1941 Percival Proctor I, G-AIWA, one of only a handful still flying in the UK and the only Mk1 model. Crop spraying was still big business in those days and two local companies showed how it was done with Miller Aerial Spraying's Grumman Ag-Cat and Sydney Garner Ltd's Hiller 12E helicopter, flown by Dave Cook. Johnny Jordan's former crop-sprayer Stearman, G-AROY, was also in evidence. Hayden Haresign, towed over from the nearby Peterborough and Spalding club by Ron Ward and Jack Wayman in their tug aircraft, very ably filled the gliding slot.

Pride of place in that year's show went to Spencer Flack's privately owned Hawker Hunter F51, G-HUNT, flown to perfection by Stefan Karwowski, who arrived over Holbeach St Johns. He had already completed two displays in Germany before returning to his base at Cranfield Aerodrome, refuelling then heading north to close the show at Fenland. B-17 *Sally B* was due to have that honour but burst a tyre on take-off at Duxford and had to abort her commitments for the day. Another pride of place must surely be given to the evocative sight of Tony Bianci's immaculate Second World War Spitfire parked, glistening in the sunlight, on the grass outside the Fenland clubhouse, from where it took off for its part in the very entertaining programme.

The new runway drainage system came into its own during the wet winter and despite a few areas that were still soggy, flying did not suffer as it had done in past winters and there were already plans afoot to remedy the dodgy areas with more drainage in the near future. These would all be in place for the Flower and Strawberry fly-ins that, together with the annual air show, would each have been running for eight years, much to the credit of the club.

Sadly the Flower fly-in drew an all-time record low with only two visitors by air, due to persistently overcast weather with a 250ft cloud base. Thankfully it was the opposite for the July Strawberry event with splendid weather and seventy-five visiting aircraft – and a fine crop of fruit!

A welcome new member in 1981 was Ray Nicholson, who formerly worked in air traffic control at West Drayton and who was now helping out in the Fenland tower. He summed up the difference as 'getting used to being actually able to see the aircraft to whom I am talking'. He would shortly be taking his Aerodrome Flight Information Service (AFIS) exam so that, as 'Fenland INFORMATION' rather than 'Fenland RADIO', flyers would receive a more comprehensive service in future. With AFIS in Ray's capable hands every weekend, he became the 'voice of Fenland' and that 'future' has now run to more than a quarter of a century – and still counting!

In 1981 the air show date was put back about three weeks to 20 September, due to the usual car parking field being under spring wheat, although the crop was actually cleared with a week to spare. Despite the club having found a 'friend in high places' in the form of Air Marshal Sir John Curtiss, who untangled loads of red tape to succeed in obtaining the first RAF participation since 1979, the weather gremlins put their oar in and the Air Ministry cancelled the Harrier at the last moment, declaring that 'the 35 knot cross wind was above limits for ejector seats when an aircraft operated at low level'. Difficulties with the weather also kept another new item, a DH Vampire, away. In danger of lacking a 'star turn', pressure from the right quarter resulted in the appearance of the BBMF Spitfire II and Hurricane II, which gave their usual sparkling display. Keeping up the wartime flavour was a second Spitfire, privately owned by the Hon. Patrick Lindsay, which beat up the airfield in fine style. Another of Patrick Lindsay's collection supported it; this time a Fieseler Storch decked out in the desert markings of the German Luftwaffe. In the stiff breeze this early example of a short take-off and landing (STOL) aeroplane hung almost stationary just a few feet above the runway and on landing came to a stop in less distance than it takes to pull up a car.

Opening the show, the strong wind gave Ian Senior a difficult time in his Pitts Special but his aerobatic skill was of the highest order, even though his cockpit canopy fractured during the display. Several of the lighter aircraft items had to be cancelled but there was still plenty to keep the crowd happy in the shape of, for example, a Piston Provost, G-AWPH, in the hands of Concorde pilot John Bradshaw, and the French naval trainer billed as the 'Cambodian Gunship' – in real life the Morane Saulnier MS 733 Alcyon, G-SHOW – that Lindsey Walton put through its paces. One of the more unusual sights was the arrival of a privately-owned Vickers Varsity, a twin-engine former RAF navigational trainer.

There was a great deal of interest, particularly in view of the wind conditions, in the slot for the Tiger Moth flown by Peter Kynsey from Redhill. The reason for this was that in order to raise funds for a local charity, Wisbech man George Wright had volunteered to fly on it, strapped to a metal framework above the upper wing! Peter would only fly the Tiger if the wind dropped below 15 knots and he had to wait until late in the afternoon before conditions were acceptable. Understandably apprehensive at take-off, George soon found his confidence and began waving merrily to the spectators. The pilot made six passes before returning him to earth again where he was mobbed by enthusiastic well-wishers.

Paying spectators reached 4,000 this year but in the face of spiralling costs, falling receipts, a recession and the constant difficulty with car parking on the cropped area, for the first time it was a struggle to make a profit and 1981 turned out to be the last time this wonderful event was held. How reminiscent of the 1930s, too!

Spring 1982 saw a new craze in flying reach Fenland when three club members (John Bugg, Andy Beale and John Keith) combined to form the Eastern Microlight Centre at the airfield. Instruction and sales were offered but the choice of the Scorpion aircraft proved to be an unforeseeable error of judgement and the centre folded after a couple of accidents involving these uninsurable machines. Meanwhile the CFI, John Jennings, was moving on towards his CPL and for a very short time Steve Davis took his place. This was not a successful arrangement but there was a sign that things were due for improvement when Nigel Newbold and Dr Carole Evans jointly negotiated the purchase of the flight centre from Doug Hopper, changing its name to Air Fenland Ltd and installing Mike Desmond as CFI.

British Airways caused something of a panic when they repossessed the Fran-Air NDB equipment but a heartfelt appeal to the people who allocated the frequency came up trumps and another 're-tread' transmitter, a Redifon G54, was donated to the club. In the capable hands of the club's electronics wizard Henry Neale it was soon up and running and NDB service resumed almost without a break.

Later in 1983, Air Fenland purchased a Cessna F172K, G-BFPH, from a member and a four-seater was once again available to rent, in addition to Don Cousins's Piper Caribbean 150, G-AREL. At this time Nigel regained his full instructor rating and Mike Desmond left to take up a post in Nigeria. By now, business was so good that John Frost was engaged as an assistant instructor but in 1984 his place, too, was taken by John Martin-Hale. By 1985, Nigel was joined by his brother, Nic, and when Nigel was forced to give up flying on medical grounds later that year, Nic had accumulated sufficient hours and rating to take over as CFI. It is sad to record that Nigel lost the fight with his illness and died in 1989.

Nic's departure in late summer 1986 and the shock discovery that BI needed an exchange engine well before the time its normal overhaul had been reached, meant that circumstances combined to bring about yet another change of ownership for the flying school. During 1985, club member Danny Ellis had reappeared after several years' absence. He progressed swiftly through his PPL and twin rating, and then he crowned a successful business career by purchasing Air Fenland. The building was redecorated, two Cessna 150s, G-ECBH and G-AWUO, were resprayed and when Mike Lomax, previously at Sibson and Booker, joined as CFI at the end of March 1987 the club was once again operating

Full circle. The modern-day hot air balloon, G-BLTA, named *James Sadler*, owned and operated by Mr K.A. Schussler, husband of a member of the Fenland Aero Club. (Mrs B. Schussler)

Aircraft

	Signature Authority Date

ANNUAL INSPECTION REPORT   MANUFACTURER ..... *COLT* ..... SERIAL No. ..... *525* ....

REGISTRATION MARK ..... *G - B LTA* .....

BALLOON NAME ..... *James Sadler* .....

TOTAL HOURS FLOWN ..... *12.10* .....

INSPECTORS DECLARATION: I certify that I inspected this balloon on ..... *28.8 1985* ..... (date) and found it to be in good condition, generally airworthy, and able to meet the approved requirements. The inspection was carried out in accordance with the approved schedule. I confirm that the entries in the log book are correct and up to date and that an approved log book is in use.

The inspection included a) flight test b) inflation c) neither (delete as applicable). List below any significant modifications carried out since the last annual inspection, or issue of C of A. (Note: any modifications not approved will invalidate the C of A) State any changes in burner, basket, or cylinders, listing serial numbers where relevant, or change of registered owner.

List any restrictions to be applied, such as change in all up weight.

SIGNED ..... *Per Lindstrand* ..... NAME (Caps) ..... *PER LINDSTRAND*

Form IR4   Technical Committee BBAC. April 1979

The annual inspection report on the hot air balloon *James Sadler* signed by Per Lindstrand, the renowned modern-day aeronaut. Dated 1985, this is 200 years after James Sadler's original first flight, which he made using a hot air balloon. (Mrs B. Schussler)

on a sound basis. Mike was joined in 1988 by his fiancée, Tina Collier, as assistant flying instructor, thus combining business with pleasure and moving the club into another settled period and – currently in the capable hands of Lee Haunch and other fresh faces – continuing to thrive in the twenty-first century.

How things have moved on in 200 years. Mr and Mrs Average can now fly anywhere in the world cheaply and with relative ease. Despite economic, safety and political factors, airshows in all shapes and sizes are as numerous and popular as ever and the number of people learning to fly, too, shows no sign of diminishing. While barnstorming in the old ways has long been 'regulated' out of business, fortunately the spirit of those pioneer flyers lives on, allowing new generations of watchers to share the feeling that flying can still be fun.

Table 1

BALLOON EVENTS IN EASTERN ENGLAND 1784–1913

DATE	ASCENT	DESCENT	AERONAUT	PASSENGERS
28/12/1784	Ely	?	Unmanned	
9/2/1785	Soham	Fordham	Unmanned	
27/5/1785	Manchester	Gainsborough	J. Sadler/unmanned	
15/7/1785	Nottingham	Edlington, Horncastle	Cracknell/unmanned	
7/10/1811	Birmingham	Heckington	J. Sadler	J. Burcham
1/11/1813	Nottingham	Pickworth	J. Sadler	None
19/8/1823	Abingdon	Parson Drove	Unmanned	
8/7/1824	Northampton	Soham	C. Green (16th ascent)	None
2/7/1825	Stamford	Thorney	C. Green (35)	Miss Stocks
8/6/1826	Boston	Bottesford	C. Green (52)	None
21/6/1826	King's Lynn	Southery	C. Green (54)	G. Green
4/8/1826	Louth	Partney, Spilsby	C. Green (56 or 57)	
30/8/1826	Boston	Manby	C. Green (59)	H. Brooke
11/9/1826	Stamford	Whittlesea	C. Green (60)	O.N. Simpson
1/10/1827	King's Lynn	?	C. Green	
14/5/1828	Boston	Gosberton	C. Green (93)	Mr Willerton
18/8/1828	Stamford	Thetford	H. Green	None
26/5/1830	Peterborough	Littleport Fen	C. Green (149)	None
6/6/1830	Peterborough	Moulton Chapel	C. Green (150)	G. Green, H. Miller
5/7/1830	Lincoln	Horncastle	C. Green (151)	G. Green
13/7/1830	Lincoln	Rothwell, Lincs	C. Green (152)	J.B. Cuttil
10/9/1834	Wisbech	?	?	
17/9/1835	London	King's Lynn	C. Green	Mr Butler
28/6/1838	Wisbech	?	?	
26/7/1841	Wisbech	Upwell	R. Gypson	G. Robins
12/8/1841	Wisbech	Bexwell	R. Gypson	R. King/M. Marshall
--/--/1840	Northampton	Local	R. Gypson	
26/8/1841	King's Lynn	Local	R. Gypson	D. Nelson
9/9/1841	King's Lynn	Dersingham	R. Gypson	A. Peek
1/10/1841	Stamford	Burghley House	R. Gypson	O.N. Simpson/ S. Sharpe/Mr Warsop Jr

DATE	ASCENT	DESCENT	AERONAUT	PASSENGERS
14/10/1841	Spalding	Holbeach Marsh	R. Gypson	None
27/10/1841	Sleaford	Quarrington	R. Gypson	None
30/4/1846	Stamford	Oxney, Peterborough	C. Green (319)	Mr Jones
17/7/1848	Lincoln	Metheringham	Mr Wadman	W. Atkinson
25/7/1848	Boston	Bennington	R. Green	T. Morton
4/9/1849	Norwich	W Newton, King's Lynn	C. Green	G. Rush
14/7/1853	King's Lynn	?	Mr Carter	?
3/8/1855	Boston	Wrangle Tofts	Capt. Chambers	None
17/7/1862	Wolverhampton	Oakham	H. Coxwell	J. Glaisher
8/7/1863	Stamford	Failed to rise	Mr Adams	None
6/9/1869	Stamford	Local	E. Jackson	None
2/9/1874	Stamford	Ryhall	B. Metcalfe	Dr Ross
10/9/1879	Boston	Freiston	J.A. Whelan	One unknown
2/8/1890	Spalding	Failed to rise	Prof Russett	None
6/8/1894	Spalding	Parachute drop	C Baldwin	None
6/8/1894	Peterborough	Coates	Joseph Simmons	Unknown
5/8/1895	Spalding	Parachute	C. Baldwin	None
5/8/1895	Peterborough	2 x Parachute	Capt. A. Orton	Miss A. Bassett (fatal)
24/6/1896	Wisbech	Parachute	C. Baldwin	None
2/8/1897	Spalding	Parachute	C. Baldwin	None
1/8/1898	Spalding	2 x Parachute	A. Gaudron	Miss A. Beaumont
7/8/1899	Spalding	2 x Parachute	A. Gaudron	Miss I. Mansfield
6/8/1900	Stamford	Parachute	Capt. A.E. Smith	None
6/8/1900	Spalding	Parachute. Abandoned (bad weather)	A. Gaudron	Miss A. Beaumont
4/8/1902	Spalding	2 x Parachute	A. Gaudron	Mr McNeil (pass)/ Mr H. Grand/ Miss Hamilton
3/8/1903	Spalding	2 x Parachute	A. Gaudron	Miss I. Mansfield/ Miss W. Mansfield
1/8/1904	Spalding	2 x Parachute	T. Emms	Miss W. Mansfield
7/8/1905	Spalding	2 x Parachute	A. Gaudron	Capt. A.E. Smith (pass)/Miss I. Cavanagh/ Mr T. Emms
30/9/1906	Paris	Fylingdales (Yorks)	Lahm/Hershey	USA
		New Holland (Hull)	Vonwiller/Cianetti	Italy
		Sandringham (Norf)	Hon. C.H. Rolls/ Col J.E. Capper	Great Britain
		Gt Walsingham (Norf)	Vaulx/Oultremont	France
17/7/1907	King's Lynn	Watlington	Lt Lampriere	None
5/8/1907	Spalding	Local	A. Gaudron	A.E. Hallas

Table 1 267

DATE	ASCENT	DESCENT	AERONAUT	PASSENGERS
1/6/1913	Farnborough	Peterborough	Lt Usbourne RN	Unknown (two balloons)
16/6/1913	Paris	Fotherby (Louth)	Pierron, No.9 (Winner)	
		Skellingthorpe (Lincoln)	Burgeois & Pulido, No.16	
		Thurlby (Bourne)	Unknown	
		Ayston (Uppingham)	Spire & Dubois	
		Marshchapel (N. Coates)	M & Mme Leblanc	

Table 2

FLYING EXHIBITION AND BARNSTORMING EVENTS IN
EASTERN ENGLAND 1912–1939

DATE	VENUE	PILOT	AEROPLANE
29/6/1912	Peterborough	W.H. Ewen	Caudron
8/7/1912	Lincoln	W.H. Ewen	Caudron
17/7/1912	Skegness	J. Brereton	Blackburn
1/8/1912	Lincoln	B.C. Hucks	Blériot
12/9/1912	Stamford	B.C. Hucks	Blériot
24/1/1913	Lincoln	B.C. Hucks	Blériot
4/5/1913	Nottingham	B.C. Hucks	Blériot
19/5/1913	Sleaford	B.C. Hucks	Blériot
29/5/1913	Peterborough	B.C. Hucks	Blériot
30/5/1913	Spalding	B.C. Hucks	Blériot
6/6/1913	Stamford	B.C. Hucks	Blériot
11/6/1913	Boston	B.C. Hucks	Blériot
13/6/1913	Louth	B.C. Hucks	Blériot
26/6/1913	Skegness	B.C. Hucks	Blériot
27/6/1913	Lincoln	E.R. Whitehouse	Handley Page E
21/8/1913	Skegness	B.C. Hucks	Blériot
25/10/1913	Sutton, Ely	J. Dagnell	'Bovril' Airship
25/6/1914	Peterborough	F.W. Goodden	Morane
7/7/1914	Hunstanton	F.P. Raynham	Avro 504
21/7/1914	March	M. Manton	Blériot
1914-1918	First World War		
8/1919	Hunstanton	Cambridge Sch. of Flying	Avro 504K G-EAHL
9/1919	Skegness	Cambridge Sch. of Flying	Avro 504K G-EAHL, DH6 G-EALT
6/1920	Peterborough	Summerfield & Co.	Avro 504K G-EADR
1921	Boston	DH/Alan Cobham	DH9C G-EAYT
5/1923	Skegness	W.A. Rollason/E. Milton	DH6
7/1923	March	Berkshire Aviation Co.	Avro 504K G-EBFV, EAKX
8/1923	Wisbech	Berkshire Aviation Co.	ditto
9/1923	Boston	Berkshire Aviation Co.	ditto
9/1923	Bourne, Crowland	Berkshire Aviation Co.	ditto

DATE	VENUE	PILOT	AEROPLANE
	Spalding, Holbeach, Long Sutton		
5/1926	King's Lynn	Lloyd Aviation Co.	Sopwith Gnu G-EAGP
5/1929	Spalding	Sir Alan Cobham	DH61 Giant Moth G-AAEV
5/1929	Boston	Sir Alan Cobham	ditto
10/1929	King's Lynn	Surrey Flying Svs	Avro 504K G-EBDP
5/1930	Horsey Toll, Peterborough	Northants Aero Club	Club air show
7/1930	Downham Mkt	Wolverhampton Av Co.	Avro 504K G-EBKR
8/1931	Horsey Toll, Peterborough	Northants Aero Club	Club air show
3/1932	King's Lynn	North-Eastern Av Co.	Spartan
7/1932	Littleport	Eastern Air Transport	DH60G Moth G-AAKM
Mid-1934	Hunstanton	Gower/Spicer	Spartan G-ABKK
8/1935	Hunstanton	Gower/Spicer	DH Fox Moth G-ADNF
Mid-1939	Skegness	Kennings Ltd	Spartan G-ABET & ABKJ

Table 3

A SELECTION OF AIR SHOWS IN EASTERN ENGLAND 1931-1939

NAD = National Aviation Day
BHAP = British Hospitals Air Pageant
BEAD = British Empire Air Display (not to be confused with 'Empire Air Day')
JAD = Jubilee Air Display

1931

8 May	SPALDING	Proctor's Farm, Wykeham. C.D. Barnard Air Tours Ltd
5 June	GRANTHAM	C.D. Barnard Air Tours Ltd
14–15 July	LINCOLN	Bracebridge Heath. C.D. Barnard Air Tours Ltd
16 July	HEACHAM	C.D. Barnard Air Tours Ltd
2 Oct.	SPALDING	Proctor's Farm, Wykeham. C.D. Barnard Air Tours Ltd

1932

15 May	SKEGNESS	Roman Bank. Skegness & East Lincs Aero Club
14 June	GRANTHAM	Spittalgate Hill, Saltersford. Cobham, NAD
15 June	BOSTON	Boardsides, Sleaford Rd. Cobham, NAD
17 June	LINCOLN	Bracebridge Heath. Cobham, NAD
12 July	SKEGNESS	Roman Bank. Cobham, NAD
19 July	PETERBOROUGH	Westwood Airfield. Cobham, NAD
20 July	HEACHAM	Stoney Hills. Cobham, NAD
31 July	SKEGNESS	Roman Bank. Skegness & East Lincs Aero Club
10 Sep.	SKEGNESS	Roman Bank. Skegness & East Lincs Aero Club

1933

19 May	PETERBOROUGH	Westwood Airfield. C.W.A. Scott, BHAP
31 May	BOSTON	Boardsides, Sleaford Rd. Cobham, NAD No.1 Tour
31 May	LINCOLN	RAF Waddington. C.W.A. Scott, BHAP
1–2 June	BOSTON	Skirbeck Grange, Sibsey Rd. C.W.A. Scott, BHAP
4 June	SKEGNESS	Roman Bank. Skegness & East Lincs Aero Club

7 June	LINCOLN	Bracebridge Heath. Cobham, NAD No.1 Tour
14 June	WISBECH	Leverington Common. Cobham, NAD No.2 Tour
21 June	SLEAFORD	Mareham Lane. Cobham, NAD No.2 Tour
2 Aug.	KING'S LYNN	Sayers Marsh, Exton Place. C.W.A. Scott, BHAP
3 Aug.	SPALDING	Proctor's Farm, Wykeham. C.W.A. Scott, BHAP
15 Aug.	SKEGNESS	Roman Bank airfield. Cobham, NAD No.2 Tour
22 Sep.	PETERBOROUGH	Westwood airfield. Cobham, NAD No.2 Tour
24 Sep.	HEACHAM	Courtyard Farm, Ringstead. Cobham, NAD No.2 Tour
26 Sep.	ELY/LITTLE DOWNHAM	Wilson's Field, Downham Rd. Cobham, NAD No.2 Tour

1934

24 May	EMPIRE AIR DAY	Including RAF Bircham Newton, Grantham, Wittering.
29 May	LOUTH	Elkington Cow Pastures. Cobham, NAD
1 Sep.	HEACHAM	Church Farm, Ringstead Rd. Cobham, NAD
11 Sep.	ELY/LITTLE DOWNHAM	Capt. Wilson's Field, Downham Rd. Cobham, NAD
17 Sep.	LINCOLN	Cobham, NAD
18 Sep.	SUTTON ON SEA (Lincs)	Brickyards. Alford Rd. Cobham, NAD

1935

24 May	EMPIRE AIR DAY	Including RAF Bircham Newton, Cranwell, Digby, Grantham, Waddington, Wittering and Sywell (civil)
19 June	OAKHAM (Rutland)	Green Lodge Farm. Campbell Black, JAD
2 July	STAMFORD	San Foin Field, Empingham Rd. Cobham, NAD (*Ferry*)
15 July	HOLBEACH/FLEET	Mays Dyke Lane, Fleet Ch End. Cobham, NAD (*Astra*)
16 July	BOURNE/MORTON	Sleaford Rd, Morton. Cobham, NAD (*Astra*)
18 July	SUTTON ON SEA (Lincs)	Brickyards, Alford Road. Cobham, NAD (*Astra*)
	LINCOLN	Campbell Black, JAD
29 July	HEACHAM	Church Fm, Stoney Hills. Cobham, NAD (*Astra*)
6 Aug.	RAMSEY/UPWOOD (Hunts)	Upwood Airfield. Cobham, NAD (*Astra*)
30 Aug.	SPALDING	Wykeham. Campbell Black, JAD
3 Sep.	WISBECH	Leverington Common. Cobham, NAD (*Ferry*)
4 Sep.	HORNCASTLE	Scholey's Field, Louth Rd. Cobham, NAD (*Ferry*)
17 Sep.	BOSTON	Boardsides, Sleaford Rd. Cobham, NAD (*Ferry*)

Table 3   273

1936

23 May	EMPIRE AIR DAY	Including RAF Bircham Newton, Digby, Grantham, Peterborough, Sutton Bridge, Waddington, Wittering and Sywell (civil)
12 April	PETERBOROUGH	Campbell Black, BEAD
19 June	SKEGNESS	Campbell Black, BEAD

1937

| 29 May | EMPIRE AIR DAY | Including RAF Bircham Newton, Digby, Grantham, North Coates Fitties, Peterborough, Sutton Bridge, Wittering, Wyton and Sywell (civil) |

1938

| 28 May | EMPIRE AIR DAY | Including RAF Bircham Newton, Cranwell, Digby, Grantham, North Coates Fitties, Peterborough, Sutton Bridge, Wittering, Wyton and Sywell (civil) |

1939

| 20 May | EMPIRE AIR DAY | Including RAF Bircham Newton, Cranwell, Hemswell, Manby, Waddington, Wittering, Wyton |

Bibliography

Various issues of UK national and provincial newspapers, including:

Abingdon Herald
Beverley Advertiser
Boston Standard
Cambridgeshire Chronicle
Daily Mail
Ely Standard
Islington Gazette
Leigh & Southend Times
Leicester Chronicle
Leicester Evening Mail
Leicester Mercury
Lincolnshire Chronicle and General Advertiser
Lincolnshire Echo
Lincolnshire Free Press
Lincolnshire Standard
Lincolnshire, Rutland & Stamford Mercury (Stamford Mercury)
Lynn Advertiser & West Norfolk Herald
Lynn News
Norfolk & Norwich Chronicle
Norfolk Chronicle & Norwich Gazette
Northamptonshire Times
Norwich Mercury
Peterborough Advertiser
Reading Mercury
Spalding Guardian
Stamford Herald & County Chronicle (The Bee)
Times, The
Wisbech Standard

Magazines and periodicals, including:

Aeroplane
Age, The
Air Clues
Air Link, journal of the Lincolnshire Aviation Society
Annual Register, The
Bystander, The
Connoisseur, The
Dublin Magazine, The
Farmer's Weekly
Flight
Gentlemen's Magazine
Hertfordshire Countryside
Illustrated London News
John Bull
L'Aerophile
Lincolnshire Life
London Gazette
Potato, The
Sketch, The

Books:

Aspin, Chris (1988) *Dizzy Heights*: Helmshome Local History Society
Bedford, John Duke of (1968) *The Flying Duchess*: MacDonald
Blake, Hodgson, Taylor (1984) *The Airfields of Lincolnshire Since 1912*: Midland Counties
Brett, R. Dallas (1928?) *History of British Aviation 1908–1914*, Vols 1&2: The Aviation Bookclub
Burge, Sqn Ldr C.G. ed. (?) *Encyclopaedia of Aviation*: Pitman & Sons Ltd
Cobham, Sir Alan, ed. Derrick Christopher. (1978) *A Time To Fly*: Shepheard-Walwyn
Cottrell, Leonard (1969) *Up In A Balloon*: (World's Work Ltd)
Coxwell, Henry (1887–9) *My Life And Balloon Experiences*, Vols 1 and 2: W.H. Allen
Cruddas, Colin (1994) *In Cobham's Company*: Cobham plc
Cruddas, Colin (2003) *Those Fabulous Flying Years*: Air-Britain
Curtis, Lettice (1985) *The Forgotten Pilots: Story of ATA 1939–1945*: Nelson & Saunders Ltd
Douglas, Sholto (1966) *Years Of Command*: Collins
Ege, Lennart (1973) *Balloons And Airships*: Blandford Press
Ellis, Paul (1980) *British Commercial Aircraft*: Janes
Gardiner, F.J. (1898) *History Of Wisbech & Neighbourhood 1848–1898*: Gardiner
Gardiner, Leslie (1984) *Lunardi*: Airlife
Gibbs-Smith, Charles (1976) *Early Flying Machines 1799–1909*: Book Club Associates
Gibbs-Smith, Charles (1984) *Ballooning*: Penguin
Gibson, Michael L. (1980) *Aviation in Northamptonshire*: Northamptonshire Library

Glaisher, James (1871) *Travels In The Air*: (Richard Bentley)

Goodrum, Alistair (1997) *Combat Reading!*: GMS Enterprises

Grey, C.G. ed. (1969 reprint) *Jane's All The World's Aircraft 1919*: David & Charles (Publishers) Ltd

Henshaw, Alex. (1980) *The Flight Of The Mew Gull*: John Murray Ltd

Hodgson, J.E. (1924) *History Of Aeronautics In Britain*: Oxford University Press

Jackson, A.J. (1959) *British Civil Aircraft Since 1919, Vols 1 to 3*: Putnam

Lewis, Peter (1970) *British Racing and Record-Breaking Aircraft*: Putnam

Lomax, Judy (1986) *Women Of The Air*: John Murray

Lucas, John (1973) *The Big Umbrella*: Elm Tree Books

Mason, Herbert M. (1973) *The Rise Of The Luftwaffe 1918–1940*: Cassell

Matthew, H.C.G. & Harrison, Brian eds (2004) *Oxford Dictionary Of National Biography*: Oxford University Press

Merton-Jones, A.C. (1976) *British Independent Airlines Since 1946, Vol 3*: ?

Minutes of the Royal Aeronautical Society 1866 et seq

Mondey, D. & Taylor, M.J.H. (1988) *The Guiness Book of Aircraft Facts and Feats*: Guiness Publishing Ltd

Murray, William (1985) *Luftwaffe*: George Allen & Unwin

Peel, Dave (1985) *British Civil Aircraft Registers Since 1919*: Midland Counties

Penrose, Harald (1969 etc) *British Aviation* Vols 1 to 5: Putnam and HMSO editions

Postma, Thijs (1980) *Fokker-Aircraft Builders to the World*: Janes

Rolt, L.T.C. (1985) *The Aeronauts*: Alan Sutton Publishing

Shores, Christopher (1983) *Air Aces*: Bison Books

Museums, Societies and Institutions:

Board of Trade (Civil Aircraft Accident Reports)

Boston Library

British Library (newspapers)

Cambridgeshire Collection

Islington Central Library

King's Lynn Library

Lincoln Central Library

Manchester Central Library

Museum of Norfolk Life

National Museum of Ireland, Dublin

Skegness Library

Spalding Gentlemen's Society

Spalding Library

Stamford Museum

Wisbech & Fenland Museum

Wisbech Library

Wolverhampton Library

Selected Internet Websites:

answers.com
bobdavenport.freeserve.co.uk
bom.gov.au
chemistry.mtm.edu
gasballon.be
metoffice.gov.uk
photovault.com
printsgeorge.com

Acknowledgements

I would like to thank the following people and organisations for their kind help over many years:

Dennis H. Ball
Maurice Buck
Brian Cocks
Paul Coulten
Colin Cruddas
Peter Green
Tony Hancock
Max Hundleby
Paul Jackson
John Ketteringham
John Knight
G. Stuart Leslie
Ray Nicholson, FISO Fenland Aero Club
Robert Nicholson (late)
Charles Parker
David Rayner
David Robinson OBE
Mike Stillingfleet, and chairman and committee of the Peterborough and Spalding Gliding
 Club
Luc Van Geyte and the Coupe Aéronautique Gordon Bennett web project team
Tony Wellband
Bill Welbourne and members of Fenland and West Norfolk Aircraft Preservation Society
Ray Wilson (late)
President and council of the Spalding Gentlemen's Society for access to its photograph collection
 and general archives
Chairman and committee of Fenland Aero Club
Editors and staff of many provincial newspapers, consulted since 1975 when I first began my
 research, but in particular the present and previous editors of the *Lincolnshire Free Press*,
Spalding Guardian and *Stamford Mercury*
Amy Rigg of The History Press for giving me the opportunity to bring this book to fruition
Crown copyright material is reproduced with the permission of the Controller of HMSO and
 the Queen's Printer for Scotland

Index

Related titles by The History Press

This detailed history examines the development of the airship, from the earliest, and often disastrous, attempts at making a dirigible balloon, up to the present. Covering this development in the UK, France, Germany, Italy and the US, John Richards delves into every aspect of the airship. The book ranges from the first fully controlled powered flight and the flight around the Eiffel Tower, the failed attempts to cross the Atlantic and the first London-Paris flight, to the many uses of the airship in times of war, including anti-submarine campaigns, convoy escorts and patrols.

ISBN: 978 0 7524 4536 6
£16.99

'I hope these new mechanic meteors will prove only playthings ... and not be converted into new engines of destruction to the human race.'
Horace Walpole, 1785.

Walpole's words were prophetic – within only eleven years the balloon had proved its worth at war and Joseph Montgolfier saw a potential use in his new invention of attacking the English at Gibraltar. The first 'air force' was founded by Napoleon Bonaparte. The American Civil, Franco-Prussian and First World Wars saw extensive use of balloons, but it was during the Second World War that the balloon saw more new and imaginative uses.

ISBN: 978 0 7524 2995 3
£16.99

If you are interested in purchasing other books published by The History Press, or in case you have difficulty finding any of our books in your local bookshop, you can also place orders directly through our website
www.thehistorypress.co.uk